Communications
in Computer and Information Science **587**

Commenced Publication in 2007
Founding and Former Series Editors:
Alfredo Cuzzocrea, Dominik Ślęzak, and Xiaokang Yang

More information about this series at http://www.springer.com/series/7899

Federico Rossi · Fabio Mavelli
Pasquale Stano · Danilo Caivano (Eds.)

Advances in Artificial Life, Evolutionary Computation and Systems Chemistry

10th Italian Workshop, WIVACE 2015
Bari, Italy, September 22–25, 2015
Revised Selected Papers

 Springer

Editors
Federico Rossi
Department of Chemistry and Biology
University of Salerno
Fisciano
Italy

Fabio Mavelli
Department of Chemistry
University of Bari
Bari
Italy

Pasquale Stano
Science Department
Roma Tre University
Roma
Italy

Danilo Caivano
Department of Informatics
University of Bari
Bari
Italy

ISSN 1865-0929   ISSN 1865-0937   (electronic)
Communications in Computer and Information Science
ISBN 978-3-319-32694-8   ISBN 978-3-319-32695-5   (eBook)
DOI 10.1007/978-3-319-32695-5

Library of Congress Control Number: 2016935956

Printed on acid-free paper

This Springer imprint is published by Springer Nature
The registered company is Springer International Publishing AG Switzerland

Preface

This volume of the Springer book series *Communications in Computer and Information Science* contains the proceedings of WIVACE 2015: the 10th Italian Workshop on Artificial Life and Evolutionary Computation, held in Bari (Italy), during September 22–25, 2015. WIVACE was first held in 2007 in Sampieri (Ragusa), as the incorporation of two previously separately run workshops (WIVA and GSICE). After the success of the first edition, the workshop was organized every year with the aim of offering a forum where different disciplines could effectively meet. The spirit of this workshop is to promote communication among single research "niches" hopefully leading to surprising "crossover" and "spill over" effects.

Events like WIVACE are generally a good opportunity for new-generation or soon-to-be scientists to get in touch with new subjects in a more relaxed, informal, and (last but not least) less expensive environment than in large-scale international conferences.

Traditionally focused on evolutionary computation, complex systems, and artificial life, since the last two editions the WIVACE community has been opened to researchers coming from experimental fields such as systems chemistry and biology, origin of life, and chemical and biological smart networks.

In this respect, WIVACE 2015 was jointly organized with a COST Action CM1304 (Emergence and Evolution of Complex Chemical Systems) meeting. The theme of this meeting, "Biomimetic Compartmentalized Chemical Systems," offered several intriguing topics to WIVACE participants and boosted the fertilization between theoretical and experimental approaches to complex dynamical systems. In fact, one of the scientific challenges nowadays is the creation of life from artificial chemical components. This means reproducing, at least, the basic aspects of life, such as the ability to reproduce, the compartmentalized nature, and the far-from-equilibrium character. Within the WIVACE community, experimental work has been harmonized in a well-established theoretical framework, which takes into account the complexity and intrinsic nonlinearity of life and can master and guide new experimental findings.

As editors, we wish to express gratitude to all the attendees of the conference and to the authors who spent time and effort to contribute to this volume. We also acknowledge the precious work of the reviewers and of the members of the Program Committee. Special thanks, finally, to the invited speakers for their very interesting and inspiring talks: Marco Dorigo from the Université Libre de Bruxelles, Belgium, Marco Gori, from the University of Siena, Italy, Ricard Solè, from the Universitat Pompeu Fabra, Barcelona, Spain, Kepa Ruiz-Mirazo, from the University of the Basque Country, Spain, Davide Scaramuzza, from the University of Zurich, Switzerland, Christodoulos Xinaris, from the research institute Mario Negri, Italy and Annette Taylor, from the University of Sheffield, UK.

The 18 papers presented were thoroughly reviewed and selected from 45 submissions. They cover the following topics: evolutionary computation, bioinspired algorithms, genetic algorithms, bioinformatics and computational biology, modelling and simulation of artificial and biological systems, complex systems, synthetic and systems biology, and systems chemistry — and they represent the most interesting contributions to the 2015 edition of WIVACE.

September 2015

Federico Rossi
Fabio Mavelli
Pasquale Stano
Danilo Caivano

Organization

WIVACE 2015 was organized in Bari (Italy) by the University of Bari (Italy) and the University of Salerno (Italy).

Program Chairs

Federico Rossi	University of Salerno, Italy
Fabio Mavelli	University of Bari, Italy
Pasquale Stano	University of Roma Tre, Italy
Danilo Caivano	University of Bari, Italy

Program Committee

Michele Amoretti	University of Parma, Italy
Lucia Ballerini	University of Edinburgh, UK
Vitoantonio Bevilacqua	University of Bari, Italy
Leonardo Bocchi	University of Firenze, Italy
Stefano Cagnoni	University of Parma, Italy
Angelo Cangelosi	University of Plymouth, UK
Timoteo Carletti	University of Namur, France
Antonio Chella	University of Palermo, Italy
Chiara Damiani	University of Milano-Bicocca, Italy
Pietro Favia	University of Bari, Italy
Alessandro Filisetti	European Centre for Living Technology, Italy
Francesco Fontanella	University of Cassino, Italy
Mario Giacobini	University of Torino, Italy
Alex Graudenzi	University of Milano-Bicocca, Italy
Roberto Marangoni	University of Pisa, Italy
Giancarlo Mauri	University of Milano-Bicocca, Italy
Alberto Moraglio	University of Birmingham, UK
Giuseppe Nicosia	University of Catania, Italy
Stefano Nolfi	Institute of Cognitive Sciences and Technologies, CNR, Italy
Gerardo Palazzo	University of Bari, Italy
Antonio Piccinno	University of Bari, Italy
Stefano Piotto	University of Salerno, Italy
Clara Pizzuti	CNR-ICAR, Italy
Andrea Roli	University of Bologna, Italy
Federico Sassi	HENESIS s.r.l., Italy
Roberto Serra	University of Modena and Reggio, Italy
Giandomenico Spezzano	ICAR-CNR, Italy
Pietro Terna	University of Torino, Italy
Andrea Tettamanzi	University of Nice Sophia Antipolis, France
Marco Villani	University of Modena and Reggio, Italy

Supported By

UNIVERSITY OF BARI, ITALY

DIPARTIMENTO
DI CHIMICA

DIPARTIMENTO
DI INFORMATICA

UNIVERSITY OF SALERNO, ITALY

Software Engineering Research & Practices

Contents

Systems Chemistry and Biology

Evolutionary Computation
and Genetic Algorithms

A Fair Comparison Between Standard PSO Versions

Roberto Ugolotti and Stefano Cagnoni[(⊠)]

Department of Information Engineering, University of Parma, Parma, Italy
cagnoni@ce.unipr.it

Abstract. Too often, when comparing a set of optimization algorithms, little effort, if any at all, is spent for finding the parameter settings which let them perform at their best on a given optimization task. Within this context, automatizing the choice of their parameter settings can be seen as a way to perform fair comparisons between optimization algorithms.

In this paper we first compare the performances of two standard PSO versions using the "standard" parameters suggested in the literature. Then, we automatically tune the parameter values of both algorithms using a meta-optimization environment, to allow the two versions to perform at their best.

As expected, results obtained by the optimized version are substantially better than those obtained with the standard settings. Moreover, they generalize well on other functions, allowing one to draw interesting conclusions regarding the PSO parameter settings that are commonly used in the literature.

Keywords: Particle swarm optimization · Meta-optimization · Automatic parameter tuning

1 Introduction

Every year, many "novel" versions of bio-inspired algorithms are presented to the scientific community, by authors who claim that they are able to easily outperform the older versions. This kind of papers has been described in [1] as "horse-race papers". In spite of their huge number, the vast majority is immediately forgotten. The main reason is that winning a race is easy if you can accommodate the rules, choose the track and select your opponents.

Comparing different algorithms is a critical task because, in addition to many other requirements, making a fair comparison implies that all algorithms must be run at their best, to avoid drawing conclusions which may be suggested by results of wrong implementations or sub-optimal parameter settings. A possible way to do so is to let an automatic parameter tuning technique tune the parameters of the algorithms under consideration to the problem at hand, avoiding any external biases such as assumptions made by the researchers themselves or use of "standard" parameter settings.

© Springer International Publishing Switzerland 2016
F. Rossi et al. (Eds.): WIVACE 2015, CCIS 587, pp. 3–14, 2016.
DOI: 10.1007/978-3-319-32695-5_1

The goal of this paper is to perform an unbiased comparison between the performance of two standard Particle Swarm Optimization (PSO [2]) versions using an automatic parameter tuning procedure that selects the settings for each algorithm to let them both perform at their best.

In the remainder of this paper we present the Standard PSO versions taken into consideration (Sect. 2) and the method used to automatically optimize their parameters (Sect. 3). Then, Sect. 4 will describe the experimental setting and the results of the comparison. Finally, in Sect. 5 we will summarize the conclusions and present some possible future extensions of this work.

2   Standard PSO Versions

PSO is a stochastic population-based metaheuristic first introduced by Kennedy and Eberhart in 1995. Since then, it has been used in a huge number of applications [3] and almost as many variants to the original algorithm have been proposed [4]. These variants mainly modify the communication among particles [5] and the position and velocity update functions [6]. To simplify the analysis of PSO variants by providing references, "Standard" PSO versions have been defined by some researchers [7].

Three Standard PSO (SPSO) versions have been presented so far: SPSO2006, SPSO2007 and SPSO2011. The first two differ only very slightly, therefore we will just compare SPSO2006 and SPSO2011.

2.1   SPSO 2006 and SPSO 2007

The goal of PSO is finding the optimum of a function within a search space $\mathbb{S}$ of dimensionality D. $\mathbb{S}$ can be considered without loss of generality as a hyperparallelepid limited by lower and upper limits for each dimension ($l_d \leq x_d \leq u_d, d = 1, \ldots, D$). To do so, a population composed of P elements ($P = 10 + int(2 \cdot \sqrt{D})$) moves across the search space $\mathbb{S}$. Each particle $P_i, i = 1, \ldots, P$ in the population is characterized by its position $x_i(t)$ and velocity $v_i(t)$, which are randomly initialized:

$$x_d(0) = U(l_d, u_d)$$
$$v_d(0) = \frac{U(l_d, u_d) - x_d(0)}{2}$$

where $U(a, b)$ is a uniform random number in $[a, b]$. The particles then update their velocity and position according to the following equations:

$$v_i(t) = w \cdot v_i(t-1) + U(0, c_1) \cdot (p_i - x_i(t-1)) + U(0, c_2) \cdot (l_i - x_i(t-1))$$
$$x_i(t) = x_i(t-1) + v_i(t) \tag{1}$$

The parameter w (inertia factor) is set to $\frac{1}{2 \cdot ln(2)} \simeq 0.721$ and $c_1 = c_2 = 0.5 + ln(2) \simeq 1.193$. The term p_i represents the best-fitness position visited so far by particle i, while l_i is the best-fitness position visited so far by any particle in particle P_i's neighborhood.

A neighborhood is the set of particles that informs a particle. Besides the original *global-best* and *local-best*, a commonly-used topology is the so-called *ring* topology, a toroidal topology according to which each particle's neighborhood is composed of itself, the previous K and the following K particles, according to an arbitrary order which is set during intialization. In SPSO2006 the neighborhood is structured according to the "adaptive random topology" (ART). In the early iterations of the optimization process, and after each iteration with no improvement of the best fitness value reached so far, the graph of the information links is updated: each particle informs itself and $K = 3$ randomly chosen particles (the same particle may be chosen several times). This means that each particle can be informed by a number of particles between 1 and P. An alternative is to compute a probability $p = 1 - (1 - \frac{1}{P})^K$ and add a link between two particles i and j only if $U_{i,j} < p$. We call this topology "probabilistic random topology" (PRT).

Equation 1 is slightly different for the SPSO2007 version: when $p_i = l_i$, the last element in the velocity update equation is removed. The other slight differences between these two versions are described in [7].

When a particle falls outside $\mathbb{S}$, its position is set as the minimum (or maximum) valid value for all coordinates and its velocity is reset to 0.

A study on convergence analysis and parameter selection of SPSO2006 using stochastic process theory has been presented in [8], where the regions of the w, c_1, c_2 space where local convergence is guaranteed have been located.

2.2   SPSO 2011

SPSO2011 differs from SPSO2006 by a few details. One is velocity initialization:

$$v_d(0) = U(l_d - x_d(0), u_d - x_d(0))$$

Another is velocity re-initialization when a particle exits the search space:

$$v_d(t+1) = -0.5 \cdot v_d(t)$$

However, the most significant one regards position and velocity updates. Let $g_i(t)$ be the center of gravity of three points: $x_i(t)$, a point "close" to $p_i(t)$, and another one close to $l_i(t)$:

$$g_i(t) = \frac{1}{3} \cdot [x_i(t) + (x_i(t) + c_1(p_i(t) - x_i(t))) + c_2(l_i(t) - x_i(t)))]$$

A new point $x_i'(t)$ is randomly selected within the hypersphere centered in $g_i(t)$ with radius $\|g_i(t) - x_i(t)\|$. Then, the particle's velocity is updated as:

$$v_i(t+1) = w \cdot v_i(t) + x_i'(t) - x_i(t) \tag{2}$$

and the new position as:

$$x_i(t+1) = w \cdot v_i(t) + x_i{}'(t)$$

These features make SPSO2011 invariant to rotation and help to extract new points from a more uniform distribution, avoiding biases which characterized the previous versions.

Other differences regard the parameters of PSO. The population size is not bound to the dimensionality of the problem and its choice is left to the developer, even if authors suggest a value of 40.

Much work has been done to derive general results on some properties of SPSO versions. Bonyadi and Michalewicz [9] studied some properties of SPSO2011, namely stability, local convergence, and sensitivity to rotation. They proved that SPSO2011 is invariant to rotation, scaling and translation of the search space and that the parameter values (they simplified the problem by fixing $c = c_1 = c_2$) for which SPSO2011 reaches stability change with the dimensionality of the problem. This is due to the fact that, while in the previous PSO versions the dimensions were all updated independently of each other, in this case the update rules involve all dimensions at the same time.

Zambrano-Bigiarini et al. [10] have applied SPSO2011 to the CEC 2013 benchmark [11], showing that SPSO2011 performs well on separable and unimodal functions but has problems with the optimization of all the composition functions and, in general, with the non-separable, asymmetrical, highly multimodal ones. A relevant property of such a PSO version is its scalability, since it can yield similar performances with different problem dimensions.

3   Meta-Optimization

Among the many possible ways to automatically select the best parameter values for a metaheuristic (MH), Meta-Optimization (see Fig. 1) is one of the most commonly used.

The block in the lower part of the image represents a traditional optimization problem: a MH, referred to as Lower-Level MH (LL-MH) optimizes a function.

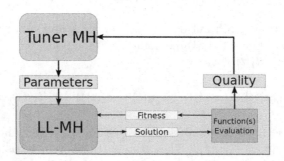

Fig. 1. Scheme of Meta-Optimization. The Tuner MH searches in the space of Lower Level MH (LL-MH) parameters. When a LL-MH configuration is generated, it is tested T times and an aggregated result is taken to be its fitness.

The Tuner MH (above in the figure) works very similarly, except for operating in the search space of the parameters of LL-MH rather than in the search space of the problem to be solved. This means that Tuner MH generates possible LL-MH configurations. For each set of parameters, an entire optimization process using LL-MH is repeated T times on the function(s) taken into consideration. An aggregated measure of the results of T (e.g., the average final fitness, or the time needed to reach a solution), called "Quality" in the image, represents the fitness value of a member of the Tuner MH population, which is run until it converges or until a certain time budget runs out.

The idea underlying Meta-Optimization was first introduced in [12], while Grefenstette [13] demonstrated the effectiveness of a meta-GA a few years later.

More recently, several methods based on the same paradigm have been proposed: REVAC (Relevance Estimation and Value Calibration [14]) is a method inspired by the Estimation of Distribution Algorithm (EDA [15]). Meissner et al. [16] used PSO to tune itself and applied the optimal parameter sets to neural network training; Pedersen [17] used Local Unimodal Sampling to tune DE and PSO parameters, obtaining good performance and discovering unexpectedly good parameter settings. ParamILS [18] performs a local search starting from a default parameter configuration, which is iteratively improved by modifying one parameter at a time; FocusedILS [19] improves ParamILS by spending more budget on the most promising areas.

A comprehensive review of Meta-Optimization methods, and automatic parameter tuning in general, can be found in [20].

3.1 Meta-Optimization Environment

SEPaT (Simple Evolutionary Parameter Tuning) is an implementation of the Meta-Optimization paradigm, introduced in [21]. In this case the algorithm used as Tuner-MH is Differential Evolution (DE [22]).

In this approach, each individual in a Tuner-MH represents a set of parameters of the LL-MH which is to be tuned. This optimizer is tested T times on a set of F functions. The resulting fitness is composed of a $2 \times F$ array of values (F averages and F standard deviations). These fitness values are then used by the tuner when its elements need to be compared. A set of parameters is better than another if it obtains better results on the majority of the F functions. In case of a tie, the winner is selected comparing the sum of Welch's t-test values over the F functions.

A problem with meta-optimizing PSO is the need to tune both numerical and nominal parameters, which are choices (e.g. PSO topology) that cannot be ordered according to any criterion and therefore are not directly searchable by a continuous optimization method like DE. The strategy we followed consists of representing each possible choice by a vector with as many real-valued elements as the options available, and selecting the option corresponding to the element having the largest value. Therefore the problem dimension is equal to: $N_{num} + \sum_{i=1}^{N_{nom}} N_c(i)$, where N_{num} is the number of numerical parameters, N_{nom} is the

Fig. 2. Encoding of PSO configurations in a tuner MH.

number of nominal parameters, and $N_c(i)$ is the number of choices available for the ith nominal parameter.

We optimized the following numerical parameters: population size, inertia factor w, c_1, c_2, and neighborhood size K; the nominal parameters were:

- Topology: the three possible choices are *ART*, *PRT*, and *ring*;
- Particles' update: the possible choices are following the SPSO2006 strategy (Eq. 1) or SPSO2011 (Eq. 2).

Therefore, the DE individuals within the tuner are encoded as in Fig. 2. The parameters of the DE-based tuner were set as: 64 individuals, 40 generations, crossover rate of 0.9, scale factor 0.5, target-to-best mutation, exponential crossover. The ranges allowed for parameters' values are presented in Table 1.

Table 1. Ranges of variability for the PSO parameter values allowed during meta-optimization. The ranges for the numerical parameters are much wider than those usually considered in the literature, to allow the tuner to "think outside the box". During meta-optimization all values are normalized in the range $[0, 1]$ and a linear scaling transformation is performed when a LL-MH is instantiated.

Parameter	Range/Values
Population size	$[4, 300]$
Inertia factor (w)	$[-0.5, 1.5]$
c_1	$[-0.5, 4.0]$
c_2	$[-0.5, 4.0]$
K	$[1, 50]$
Topology	$\{ART, PRT, ring\}$
Update	$\{SPSO2006, SPSO2011\}$

4   Experimental Results

4.1   Standard Versions

We first present the results of a direct comparison between SPSO2006 and SPSO2011. The parameter values used by the basic versions are the ones suggested in [7] and in [10] and are the same for the two versions: population size of 40, adaptive random topology with $K = 3$, $c_1 = c_2 = 0.5 + ln(2)$ and $w = \frac{1}{2 \cdot ln(2)}$.

Table 2 presents the results obtained by SPSO2006 and SPSO2011 over the 28 functions that compose the CEC 2013 benchmark [11] in 10, 30, and 50 dimensions. For each PSO version and each function, 51 independent runs were performed with a termination criterion of $1000 \cdot D$ fitness evaluations. Median final fitness and standard deviations are reported in the tables. The last column of each group indicates which version performed better (if any) according to the results of Wilcoxon signed-rank tests with a significance value of 0.01.

In 10 dimensions, SPSO2011 performs better than SPSO2006 on 13 functions, a tie occurs in 9 cases and SPSO2006 is better 6 times. These results are generally confirmed when increasing the dimensionality to 30 (SPSO2011 wins 11 times, SPSO2006 6 times) and 50 (SPSO2011 wins 13 times, SPSO2006 6 times).

Usually SPSO2006 performs better on separable functions ($f5$, $f11$, $f22$) because it updates each dimension independently of the others and is therefore more suitable for this kind of problems. The same happens for non-rotated functions ($f14$, $f17$). In the other situations (multimodal, rotated, non-separable functions) SPSO2011 usually obtains better results. Nevertheless, both versions are unable to solve the majority of the functions.

4.2 Meta-Optimization

The next step was the meta-optimization of the parameters. We divided the (10-dimensional versions of) CEC 2013 functions into a training set and a test set. The training set included 7 functions, the remaining 21 composed the test set. To avoid biases favoring any of the two versions, the training set included three functions on which the two SPSO versions yield the same results ($f1$, $f3$, $f8$), two in which SPSO2006 performs better than SPSO2011 ($f11$, $f21$) and two for which the latter was better ($f10$, $f23$). The functions in this set were chosen such that they covered a wide range of function properties, like unimodal/multimodal, separable/non separable, symmetrical/asymmetrical and rotated/non rotated.

The set of PSO parameters obtained by SEPaT are presented in Table 3. The most striking difference with the parameter values suggested for SPSOxx regards the population size. The number of elements in the population was already suspected of being responsible for bad performance of PSO [10], implying that 40 was too small a number of elements. Nevertheless, K is proportionally lower if one considers the ratio between number of elements in the population and the neighborhood size; this suggests that the best strategy for PSO is to use many particles with little communication between one another. Considering this, strategies like niching and sub-swarming seem to be beneficial.

Interestingly, although SPSO2011 performed better using the standard parameters, SEPaT chose to update velocity and position of the particles according to SPSO2006. This choice needs to be analyzed more in depth, but it suggests

Table 2. Summary of the results of the basic versions in 10, 30, and 50 dimensions. The last column of each group indicates which version is better according to the Wilcoxon signed-rank test ($p < 0.01$) for each problem dimension.

Fun	Goal	10 SPSO2006 Median	Std	Best	10 SPSO2011 Median	Std	Best	30 SPSO2006 Median	Std	Best	30 SPSO2011 Median	Std	Best	50 SPSO2006 Median	Std	Best	50 SPSO2011 Median	Std	Best
f1	-1.400e+03	-1.400e+03	0.000e+00	-	-1.400e+03	0.000e+00	-	-1.400e+03	0.000e+00	-	-1.400e+03	0.000e+00	-	-1.400e+03	0.000e+00	-	-1.400e+03	0.000e+00	-
f2	-1.300e+03	1.006e+05	1.165e+05	2011	1.457e+04	1.694e+04	2011	1.006e+05	1.165e+05	2011	1.457e+04	1.694e+04	2011	9.109e+05	4.504e+05	2011	3.794e+05	9.345e+04	2011
f3	-1.200e+03	-9.851e+02	7.447e+05	-	1.108e+03	5.786e+05	-	-9.851e+02	7.447e+05	-	1.108e+03	5.786e+05	-	4.799e+07	1.228e+08	-	8.490e+07	1.266e+08	-
f4	-1.100e+03	6.998e+02	9.922e+02	2011	1.896e+02	7.345e+02	2011	6.998e+02	9.922e+02	2011	1.896e+02	7.345e+02	2011	3.860e+03	1.236e+03	2011	1.055e+03	6.849e+02	2011
f5	-1.000e+03	-1.000e+03	0.000e+00	2006	-1.000e+03	7.033e-05	2006	-1.000e+03	0.000e+00	2006	-1.000e+03	7.033e-05	2006	-1.000e+03	0.000e+00	2006	-1.000e+03	6.672e-05	2006
f6	-9.000e+02	-8.902e+02	3.977e+00	2006	-8.902e+02	3.571e+00	2006	-8.902e+02	3.977e+00	2006	-8.902e+02	3.571e+00	2006	-8.566e+02	1.307e+01	2006	-8.566e+02	1.967e+01	-
f7	-8.000e+02	-7.999e+02	1.300e+00	-	-7.996e+02	1.503e+00	-	-7.999e+02	1.300e+00	-	-7.996e+02	1.503e+00	-	-7.408e+02	1.544e+01	-	-7.536e+02	1.040e+01	2011
f8	-7.000e+02	-6.797e+02	6.511e-02	-	-6.797e+02	8.072e-02	-	-6.797e+02	6.511e-02	-	-6.797e+02	8.072e-02	-	-6.789e+02	4.315e-02	-	-6.789e+02	5.378e-02	-
f9	-6.000e+02	-5.978e+02	1.254e+00	-	-5.979e+02	9.012e-01	-	-5.978e+02	1.254e+00	-	-5.979e+02	9.012e-01	-	-5.400e+02	3.026e+00	2011	-5.531e+02	9.097e+00	2011
f10	-5.000e+02	-4.999e+02	5.212e-02	2011	-4.999e+02	3.619e-02	2011	-4.999e+02	5.212e-02	2011	-4.999e+02	3.619e-02	2011	-4.998e+02	1.020e-01	2006	-4.998e+02	1.402e-01	2006
f11	-4.000e+02	-3.970e+02	1.824e+00	-	-3.952e+02	1.508e+00	2006	-3.970e+02	1.824e+00	-	-3.952e+02	1.508e+00	2006	-2.856e+02	2.661e+01	2006	-2.518e+02	2.901e+01	2006
f12	-3.000e+02	-2.909e+02	4.331e+00	2011	-2.956e+02	1.850e+00	2011	-2.909e+02	4.331e+00	2011	-2.956e+02	1.850e+00	2011	-1.052e+02	3.657e+01	2011	-1.687e+02	1.977e+01	2011
f13	-2.000e+02	-1.908e+02	4.702e+00	2011	-1.945e+02	3.019e+00	2011	-1.908e+02	4.702e+00	2011	-1.945e+02	3.019e+00	2011	7.956e+01	4.152e+01	2011	5.080e+01	3.182e+01	2011
f14	-1.000e+02	6.047e+01	1.868e+02	2006	5.378e+02	1.926e+02	2006	6.047e+01	1.868e+02	2006	5.378e+02	1.926e+02	2006	5.742e+03	1.154e+03	2006	8.474e+03	6.299e+02	2006
f15	1.000e+02	9.261e+02	2.604e+02	2011	6.200e+02	1.891e+02	2011	9.261e+02	2.604e+02	2011	6.200e+02	1.891e+02	2011	1.078e+04	5.318e+02	2011	9.250e+03	6.605e+02	2011
f16	2.000e+02	2.008e+02	2.058e-01	2011	2.007e+02	1.993e-01	2011	2.008e+02	2.058e-01	2011	2.007e+02	1.993e-01	2011	2.026e+02	2.809e-01	2011	2.021e+02	3.045e-01	2011
f17	3.000e+02	3.169e+02	3.034e+00	-	3.173e+02	2.317e+00	-	3.169e+02	3.034e+00	-	3.173e+02	2.317e+00	-	4.540e+02	2.927e+01	2006	5.195e+02	3.207e+01	2006
f18	4.000e+02	4.244e+02	5.182e+00	2011	4.183e+02	2.805e+00	2011	4.244e+02	5.182e+00	2011	4.183e+02	2.805e+00	2011	7.108e+02	2.426e+01	2011	6.364e+02	2.368e+01	2011
f19	5.000e+02	5.008e+02	2.767e-01	-	5.008e+02	1.451e-01	-	5.008e+02	2.767e-01	-	5.008e+02	1.451e-01	-	5.099e+02	2.964e+00	-	5.109e+02	3.468e+00	-
f20	6.000e+02	6.025e+02	4.210e-01	-	6.025e+02	4.932e-01	-	6.025e+02	4.210e-01	-	6.025e+02	4.932e-01	-	6.213e+02	1.203e+00	-	6.197e+02	6.790e-01	2011
f21	7.000e+02	1.100e+03	0.000e+00	2006	1.268e+03	2.940e+02	2006	1.100e+03	0.000e+00	2006	1.268e+03	2.940e+02	2006	1.822e+03	3.528e+02	2006	1.822e+03	2.083e+02	-
f22	8.000e+02	9.684e+02	1.477e+02	2006	1.245e+03	2.170e+02	2011	9.684e+02	1.477e+02	2006	1.245e+03	2.170e+02	2011	6.534e+03	1.269e+03	2006	8.927e+03	9.952e+02	2006
f23	9.000e+02	1.643e+03	2.155e+02	2011	1.201e+03	1.024e+01	2011	1.643e+03	2.155e+02	2011	1.201e+03	1.024e+01	2011	1.177e+04	8.198e+02	2011	1.038e+04	9.322e+02	2011
f24	1.000e+03	1.210e+03	3.946e+00	2011	1.301e+03	1.024e+01	2011	1.210e+03	3.946e+00	2011	1.301e+03	1.024e+01	2011	1.300e+03	1.260e+01	2011	1.290e+03	1.163e+01	2011
f25	1.100e+03	1.307e+03	4.932e+00	2011	1.308e+03	2.302e+00	2011	1.307e+03	4.932e+00	2011	1.308e+03	2.302e+00	2011	1.437e+03	1.698e+01	2011	1.441e+03	1.140e+01	-
f26	1.200e+03	1.400e+03	4.584e+01	2011	1.400e+03	4.214e+01	2011	1.400e+03	4.584e+01	2011	1.400e+03	4.214e+01	2011	1.623e+03	1.123e+02	2011	1.576e+03	8.088e+01	-
f27	1.300e+03	1.700e+03	9.884e+01	2011	1.601e+03	3.729e+01	2011	1.700e+03	9.884e+01	2011	1.601e+03	3.729e+01	2011	2.810e+03	2.469e+02	2011	2.543e+03	9.482e+01	2011
f28	1.400e+03	1.700e+03	5.064e+01	-	1.700e+03	3.430e+01	-	1.700e+03	5.064e+01	-	1.700e+03	3.430e+01	-	1.800e+03	9.946e+02	-	1.800e+03	1.097e+03	-

Table 3. PSO parameter values found by the meta-optimization process.

Parameter	Value
Population size	246
Inertia factor (w)	0.687378
c_1	0.246448
c_2	0.701503
K	6
Topology	ART
Update	$SPSO2006$

that a correct parameter setting may overturn the conclusions derived from the comparison of two algorithms based on the standard parameter values.

Regarding the other numerical parameters, the inertia factor is close to the values commonly used in the literature, while c_1 and c_2 tend to be smaller than the ones usually suggested. Moreover, c_2 is larger than c_1, improving the ability of the particles to communicate with one another, rather than following their own paths. The combination of these parameters is not in the range of local convergence estimated in [8]: this is a further suggestion that the exploration ability is more important than the exploitation phase.

Table 4 shows the results obtained by the tuned version on the 28 functions of the CEC 2013 benchmark. For each of the three problem dimensions considered (10, 30, and 50), the table shows the median and the standard deviation of the final fitness. The last column of each group shows, for the functions which were not considered during the tuning phase, which PSO version obtained the best results according to the Wilcoxon signed-rank test.

It can be observed that the tuned version (indicated by TPSO) is able to improve the results of the standard versions on the majority of the functions, including those which were not within the ones that were tuned. This supremacy is also confirmed, and in some cases improved, when the dimensionality of the problem increases to 30 and 50, showing that our approach is able to find parameter sets that generalize well. In particular, the standard versions perform comparably on the unimodal functions, while the tuned version is constantly better on the multimodal and composition ones.

The plots in Fig. 3 show in which way the tuned version outperforms the standard ones. The standard versions start faster thanks to the smaller population but, after a while, they cannot improve their results any further, while the tuned version constantly keeps finding better solutions without getting stuck into local minima. The fast convergence of SPSO2011 is also witnessed by the results on the CEC 2013 competition [23]: here PSO was one of the best-performing algorithms when a very low number of function evaluations was considered, and one of the worst after the entire evaluation budget was consumed.

Table 4. Summary of the results obtained by the tuned version in 10, 30, and 50 dimensions. For each dimensionality, the last column indicates (only for the functions not included in the training set, indicated with T) which version is better between the tuned version (TPSO) and the basic versions presented in Table 2 according to the Wilcoxon signed-rank test (p < 0.01).

Fun	Goal	10			30			50		
		Median	Std	Best	Median	Std	Best	Median	Std	Best
f1	-1.400e+03	-1.400e+03	0.000e+00	T	-1.400e+03	0.000e+00	T	-1.400e+03	0.000e+00	T
f2	-1.300e+03	1.381e+04	2.074e+04	2011, TPSO	4.252e+05	1.698e+05	2011	7.115e+05	1.839e+05	2011, TPSO
f3	-1.200e+03	-1.198e+03	3.344e+03	T	4.935e+06	9.344e+06	T	2.444e+07	4.826e+07	T
f4	-1.100e+03	1.350e+03	1.241e+03	2011	9.822e+03	2.232e+03	2006, 2011	1.317e+04	3.138e+03	-
f5	-1.000e+03	-1.000e+03	3.367e-32	2006	-1.000e+03	0.000e+00	2006, TPSO	-1.000e+03	0.000e+00	2006, TPSO
f6	-9.000e+02	-8.902e+02	2.275e+00	2006, TPSO	-8.852e+02	2.565e+01	2006, 2011	-8.565e+02	1.353e+01	-
f7	-8.000e+02	-8.000e+02	1.764e-01	TPSO	-7.868e+02	8.343e+00	TPSO	-7.671e+02	6.738e+00	TPSO
f8	-7.000e+02	-6.797e+02	7.069e-02	T	-6.791e+02	6.042e-02	T	-6.789e+02	5.456e-02	T
f9	-6.000e+02	-5.987e+02	7.351e-01	TPSO	-5.895e+02	2.535e+00	TPSO	-5.740e+02	2.723e+00	TPSO
f10	-5.000e+02	-5.000e+02	2.186e-02	T	-4.999e+02	6.830e-02	T	-4.998e+02	8.464e-02	T
f11	-4.000e+02	-3.980e+02	1.430e+00	T	-3.791e+02	7.181e+00	T	-3.413e+02	1.306e+01	T
f12	-3.000e+02	-2.970e+02	1.740e+00	TPSO	-2.781e+02	6.232e+00	TPSO	-2.294e+02	1.304e+01	TPSO
f13	-2.000e+02	-1.961e+02	2.023e+00	TPSO	-1.392e+02	1.867e+01	TPSO	-1.958e+01	2.867e+01	TPSO
f14	-1.000e+02	5.815e+01	1.389e+02	2006, TPSO	2.558e+03	5.798e+02	2006, TPSO	5.240e+03	6.175e+02	TPSO
f15	1.000e+02	4.160e+02	1.298e+02	TPSO	3.024e+03	4.923e+02	TPSO	6.607e+03	8.783e+02	TPSO
f16	2.000e+02	2.007e+02	1.584e-01	2011, TPSO	2.016e+02	2.572e-01	2011, TPSO	2.021e+02	2.533e-01	2011, TPSO
f17	3.000e+02	3.129e+02	1.216e+00	TPSO	3.474e+02	4.258e+00	TPSO	3.931e+02	9.819e+00	TPSO
f18	4.000e+02	4.167e+02	2.596e+00	TPSO	4.661e+02	2.239e+01	TPSO	4.956e+02	1.445e+01	TPSO
f19	5.000e+02	5.006e+02	1.438e-01	TPSO	5.028e+02	6.572e-01	TPSO	5.061e+02	1.424e+00	TPSO
f20	6.000e+02	6.021e+02	5.517e-01	-	6.103e+02	1.236e+00	2011, TPSO	6.185e+02	1.011e+00	TPSO
f21	7.000e+02	1.100e+03	0.000e+00	T	1.000e+03	6.274e+01	T	1.822e+03	1.426e+02	T
f22	8.000e+02	8.661e+02	1.028e+02	TPSO	2.761e+03	5.915e+02	TPSO	5.674e+03	8.667e+02	TPSO
f23	9.000e+02	9.865e+02	1.550e+02	T	3.807e+03	5.605e+02	T	7.855e+03	9.868e+02	T
f24	1.000e+03	1.200e+03	2.189e+01	2011, TPSO	1.218e+03	6.402e+00	TPSO	1.258e+03	1.131e+01	TPSO
f25	1.100e+03	1.300e+03	1.301e+01	TPSO	1.360e+03	7.446e+00	TPSO	1.424e+03	1.412e+01	TPSO
f26	1.200e+03	1.304e+03	3.324e+01	2011, TPSO	1.400e+03	3.620e+01	2011, TPSO	1.540e+03	7.138e+01	TPSO
f27	1.300e+03	1.600e+03	3.188e+01	TPSO	1.737e+03	5.664e+01	TPSO	2.287e+03	9.394e+01	TPSO
f28	1.400e+03	1.700e+03	2.773e+01	-	1.700e+03	0.000e+00	-	1.800e+03	0.000e+00	-

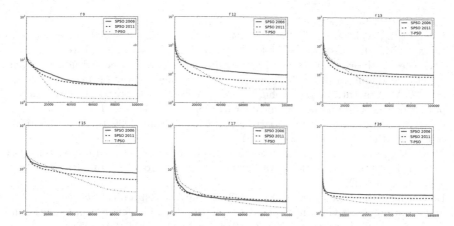

Fig. 3. Some examples in which the tuned PSO performs better than the standard version (in 10 dimensions).

5   Conclusions and Future Work

In this work, we compared two standard versions of Particle Swarm Optimization, SPSO2006 and SPSO2011. First, we compared them using standard parameter settings. Under these conditions, this comparison showed a slight supremacy of SPSO2011.

Then, we used a meta-optimization environment to tune their parameters and to select between their update strategies, and we were able to systematically improve PSO performance also on unseen functions. Regarding PSO parameter settings, the main conclusion is that, especially for multi-modal functions, PSO needs a larger population than suggested by the authors of the standard versions. Regarding topology, in the tuned version this large population is sparsely connected, which suggests that approaches like sub-swarms and niching could improve PSO's performance.

Since the results we obtained are better than for standard settings of standard versions, we suggest that researchers use the parameters reported in Table 3 when comparing a novel modification to the standard algorithm, especially when using the same popular collection of functions as benchmark.

References

1. Johnson, D.S.: A theoretician's guide to the experimental analysis of algorithms. In: Goldwasser, M.H., Johnson, D.S., McGeoch, C.C. (eds.) Data Structures, Near Neighbor Searches, and Methodology: Fifth and Sixth DIMACS Implementation Challenges, vol. 59, pp. 215–250. American Mathematical Society, Providence (2002)
2. Kennedy, J., Eberhart, R.: Particle swarm optimization. In: IEEE International Conference on Neural Networks, ICNN 1995, vol. 4, pp. 1942–1948 (1995)
3. Poli, R.: Analysis of the publications on the applications of Particle Swarm Optimisation. J. Artif. Evol. Appl. **2008**, 1–10 (2008)

4. Clerc, M.: Particle Swarm Optimization. Wiley, New York (2010)
5. Vazquez, J., Valdez, F., Melin, P.: Comparative study of social network structures in PSO. In: Castillo, O., Melin, P., Pedrycz, W., Kacprzyk, J. (eds.) Recent Advances on Hybrid Approaches for Designing Intelligent Systems. Studies in Computational Intelligence, vol. 547, pp. 239–254. Springer, Basel (2014)
6. Banks, A., Vincent, J., Anyakoha, C.: A review of particle swarm optimization. part I: background and development. Nat. Comput. **6**(4), 467–484 (2007)
7. Clerc, M.: Standard particle swarm optimisation from 2006 to 2011. Technical report, Particle Swarm Central (2011)
8. Jiang, M., Luo, Y., Yang, S.: Stochastic convergence analysis and parameter selection of the standard particle swarm optimization algorithm. Inf. Process. Lett. **102**(1), 8–16 (2007)
9. Bonyadi, M.R., Michalewicz, Z.: SPSO 2011: analysis of stability; local convergence; and rotation sensitivity. In: Proceedings of the 2014 Conference on Genetic and Evolutionary Computation. GECCO 2014, pp. 9–16. ACM (2014)
10. Zambrano-Bigiarini, M., Clerc, M., Rojas, R.: Standard particle swarm optimisation 2011 at CEC-2013: a baseline for future pso improvements. In: IEEE Congress on Evolutionary Computation, CEC 2013, pp. 2337–2344, June 2013
11. Liang, J., Qu, B., Suganthan, P.: Problem definitions and evaluation criteria for the CEC 2013 special session on real-parameter optimization. Technical report 201212, Computational Intelligence Laboratory, Zhengzhou University (2013)
12. Mercer, R., Sampson, J.: Adaptive search using a reproductive metaplan. Kybernetes **7**, 215–228 (1978)
13. Grefenstette, J.: Optimization of control parameters for genetic algorithms. IEEE Trans. Syst. Man Cybern. **16**(1), 122–128 (1986)
14. Nannen, V., Eiben, A.E.: Relevance estimation and value calibration of evolutionary algorithm parameters. In: International Joint Conference on Artifical Intelligence, IJCAI 2007, pp. 975–980 (2007)
15. Larrañaga, P., Lozano, J.A.: Estimation of Distribution Algorithms: A New Tool for Evolutionary Computation. Kluwer Academic Publishers, Norwell (2001)
16. Meissner, M., Schmuker, M., Schneider, G.: Optimized Particle Swarm Optimization (OPSO) and its application to artificial neural network training. BMC Bioinform. **7**, 125 (2006)
17. Pedersen, M.E.H.: Tuning and simplifying heuristical optimization. Master's thesis, University of Southampton (2010)
18. Hutter, F., Hoos, H.H., Leyton-Brown, K., Stützle, T.: ParamILS: an automatic algorithm configuration framework. J. Artif. Intell. Res. **36**(1), 267–306 (2009)
19. Hutter, F., Hoos, H.H., Stützle, T.: Automatic algorithm configuration based on local search. In: National Conference on Artificial Intelligence, AAAI 2007, vol. 7, pp. 1152–1157 (2007)
20. Eiben, A.E., Smit, S.K.: Parameter tuning for configuring and analyzing evolutionary algorithms. Swarm Evol. Comput. **1**(1), 19–31 (2011)
21. Ugolotti, R., Nashed, Y.S.G., Mesejo, P., Cagnoni, S.: Algorithm configuration using GPU-based metaheuristics. In: Genetic and Evolutionary Computation Conference (GECCO) Companion Proceedings, pp. 221–222. ACM (2013)
22. Storn, R., Price, K.: Differential evolution- a simple and efficient adaptive scheme for global optimization over continuous spaces. Technical report, International Computer Science Institute (1995)
23. Loshchilov, I., Stützle, T., Liao, T.: Ranking results of CEC 2013 special session & competition on real-parameter single objective optimization

On the Robustness of the Detection of Relevant Sets in Complex Dynamical Systems

Marco Villani[1], Pietro Carra[2], Andrea Roli[3(✉)], Alessandro Filisetti[4,5], and Roberto Serra[1]

[1] Department of Physics, Informatics and Mathematics, Università di Modena e Reggio Emilia & European Centre for Living Technology, Venice, Italy
[2] Department of Physics, Informatics and Mathematics, Università di Modena e Reggio Emilia, Modena, Italy
[3] Department of Computer Science and Engineering, Alma Mater Studiorum Università di Bologna, Bologna, Italy
andrea.roli@unibo.it
[4] Explora, s.r.l., Rome, Italy
[5] European Centre for Living Technology, Venice, Italy

Abstract. The identification of system's parts that rule its dynamics and the understanding of its dynamical organisation is a paramount objective in the analysis of complex systems. In previous work we have proposed the *Dynamical Cluster Index method*, which is based on information-theoretical measures. This method makes it possible to identify the components of a complex system that are relevant for its dynamics as well as their relation in terms of information flow. Complex systems' organisation is often characterised by intertwined components. The detection of such dynamical structures is a prerequisite for inferring the hierarchical organisation of the system. The method relies on a ranking based on a statistical index, which depends on a reference system (the *homogeneous system*) generated according to a parametrised sampling procedure. In this paper we address the issue of assessing the robustness of the method against the homogeneous system generation model. The results show that the method is robust and can be reliably applied to the analysis of data from complex system dynamics in general settings, without requiring particular hypotheses.

Keywords: Information theory · Mutual information · Information integration · Dynamical cluster index · Dynamical system · Boolean networks

1   Introduction

Complex systems often show forms of tangled organisation, characterised by the intertwined relation of their parts. When modelling a complex system it is common to associate variables to its atomic components and the actual dynamics of the system can be represented by the interplay among some of its groups

© Springer International Publishing Switzerland 2016
F. Rossi et al. (Eds.): WIVACE 2015, CCIS 587, pp. 15–28, 2016.
DOI: 10.1007/978-3-319-32695-5_2

of variables, which we may call *relevant subsets*. The identification of relevant subsets in a complex system is a key issue in contexts as diverse as biology, neuroscience, environmental studies, big data analysis, economics and robotics, just to mention some. In previous works, we have proposed a method based on information-theoretical measures that aims at identifying those relevant subsets (heareafter referred to as RSs). The technique is named the *Dynamical Cluster Index method* (DCI) and has been shown to effectively capture the dynamical organisation of complex systems of several kinds, such as genetic networks and chemical reactions [1,14]. This method has been shown to be superior to classical correlation-based techniques and can be applied to dynamical systems non necessarily in stationary states. Recently, we have proposed an operational definition of RS in terms of a filtering algorithm and suggested the use of transfer entropy for the assessment of the information flow among RSs [2]. The overall method developed makes it possible to identify the components of a complex system that are relevant for its dynamics as well as their relation in terms of information flow. The detection of such dynamical structure is a prerequisite for inferring a likely hierarchical organisation of the system. The identification of the RSs relies on a ranking based on a statistical index, which depends on a reference system. This system should provide a baseline for the assessment of the results and it is usually generated by means of a stochastic model, so as to both capture the main statistical properties of the complex system to be analysed and provide a reference *homogeneous system*, i.e. a system without RSs. A subtle yet crucial point concerns the robustness and significance of the results attained by the application of the DCI method against the fluctuations induced by the sampling of the homogeneous system.

In this paper we address this issue and show that the results attained by the DCI method are valid in general and do not depend upon the homogeneous system instances generated. We first briefly recall the basic notions and provide an overview of the DCI method in Sect. 2. In Sect. 3 we succinctly illustrate the sieving procedure used for identifying the RSs and their exchange of information. The robustness of the results is discussed in Sect. 4, where we present the results of a thorough statistical analysis. Finally, we discuss further improvements of the method and we conclude with Sect. 6.

2   The Dynamical Cluster Index Method

The DCI method has been introduced in previous work and the interested reader can find details in [1,12,13]. For the sake of completeness, we provide a brief summary of the main notions and the method itself. Let us consider a system modelled with a set U of n variables ranging in finite and discrete value domains. The cluster index of a subset S of variables in U, $S \subset U$, as defined by Tononi et al. [10], estimates the ratio between the amount of information integration among the variables in S and the amount of integration between S and U. These quantities are based on the Shannon entropy of both the single elements and sets of elements in U. The entropy of an element x_i is defined as:

$$H(x_i) = - \sum_{v \in V_i} p(v) \; log \; p(v) \tag{1}$$

where V_i is the set of the possible values of x_i and $p(v)$ the probability of occurrence of symbol v.

In this work, we deal with observational data, therefore probabilities are estimated by means of relative frequencies. The cluster index $C(S)$ of a set S of k elements is defined as the ratio between the integration $I(S)$ of S and the mutual information between S and the rest of the system $U - S$. The integration of S is defined as:

$$I(S) = \sum_{x \in S} H(x) - H(S) \tag{2}$$

$I(S)$ represents the deviation from statistical independence of the k elements in S. The mutual information $M(S; U - S)$ is defined as:

$$M(S; U - S) \equiv H(S) - H(S|U - S) = H(S) + H(U - S) - H(S, U - S) \tag{3}$$

where $H(A|B)$ is the conditional entropy and $H(A, B)$ the joint entropy. Note that $H(A)$ denotes the entropy of the set A. Finally, the cluster index $C(S)$ is defined as:

$$C(S) = \frac{I(S)}{M(S; U - S)} \tag{4}$$

Since C is defined as a ratio, it is undefined in all those cases where $M(S; U - S)$ vanishes. In this case, the subset S is statistically independent from the rest of the system and it has to be analyzed separately. As $C(S)$ scales with the size of S, cluster index values of systems of different size need to be normalized. To this aim, a reference system is defined, i.e., the homogeneous system U_h, that keep some statistical properties of the original system but does not contain clusters. There are several ways to generate U_h, which will be discussed in Sect. 4. In general, for each subsystem size of U_h the average integration I_h and the average mutual information M_h are computed. The cluster index value of S is normalized by means of the appropriate normalization constant:

$$C'(S) = \frac{I(S)}{\langle I_h \rangle} / \frac{M(S; U - S)}{\langle M_h \rangle} \tag{5}$$

Furthermore, to assess the significance of the differences observed in the cluster index values, a statistical index T_c is computed:

$$T_c(S) = \frac{C'(S) - \langle C'_h \rangle}{\sigma(C'_h)} = \frac{C(S) - \langle C_h \rangle}{\sigma(C_h)} \tag{6}$$

where $\langle C_h \rangle$, $\sigma(C_h)$ and $\langle C'_h \rangle$ and $\sigma(C'_h)$ are the average and the standard deviation of the population of cluster indices and normalized cluster indices with the same size of S from the homogeneous system.

The search for RSs of a dynamical system by means of the DCI requires first the collection of observations of the values of the variables at different instants.

As the DCI is computed on the basis of symbol frequencies, in principle it is not required to have a time series but just a collection of snapshots of the system variables. However, the results discussed in this paper have been obtained by analysing time series of discrete-time discrete-state systems. In order to find the RSs, in principle all the possible subsets of system variables should be considered and their DCI computed. In practice, this procedure is feasible only for small-size subsystems in a reasonable amount of time. Therefore, heuristics are required to address the study of large-size systems.

3   Relevant Subsets

The list of candidate RSs (CRSs) can be ranked according to the significance of their DCI. In many cases this analysis might return a huge list of entangled sets, so that a direct inspection is required for explaining their relevance. To this aim, we have introduced a DCI analysis post-processing sieving algorithm to reduce the overall number of CRSs to manually tackle [2]. A main limitation might be owing to the fact that if a CRS A is a proper subset of CRS B, then only the subset with the higher DCI is maintained between the two. Thus, only disjoint or partially overlapping CRSs are retained: the used assumption implies that the remaining CRSs are not further decomposable, forming in such a way the "building blocks" of the dynamical organisation of the system. The sieving algorithm enables us to provide a precise definition of RSs, avoiding the fuzzyness typical of the definitions of clusters in general. Of course, this operational definition is based on some assumptions, which might limit the outcome of a DCI-based analysis. The main assumption is that the ranking of the CRSs depends upon the homogeneous system: this issue will be thoroughly discussed in Sect. 4.

3.1   Information Flow Among RSs

The transfer entropy (TE) has been introduced by Schreiber [6] as a measure to quantify information transfer between systems evolving in time. Let X and Y be two random variables representing the state transition of two stochastic or deterministic systems. Let x_t and y_t be the values respectively of X and Y at time t. Let also suppose that the systems are Markovian processes of order 1, i.e. $p(x_{t+1}|x_t, x_{t-1}, x_{t-2}, \ldots) = p(x_{t+1}|x_t)$.[1] The transfer entropy $T_{X \to Y}$ quantifies the amount of information available from knowing y_t on x_{t+1} and is defined as follows:

$$T_{Y \to X} = \sum_{X,Y} p(x_{t+1}, x_l, y_t) \; log \; \frac{p(x_{t+1}|x_t, y_t)}{p(x_{t+1}|x_t)}, \tag{7}$$

We note that the temporal dependency is not necessarily of unitary lag, i.e. $t - 1 \to t$. For a complete assessment of the statistical dependency of X on Y one should sum over $t - 1, t - 2, \ldots, t - k$, where k is the order of the Markovian

---

[1] We will further discuss the consequences of this hypothesis.

Table 1. The update rules of the boolean networks discussed in the text. Random(0.5) denotes a Bernoulli distribution with probability 0.5.

Node	Node Rule				
	Case 1	Case 2	Case 3	Case 4	Case 5
N01	Random(0.5)	Random(0.5)	Random(0.5)	Random(0.5)	Random(0.5)
N02	Random(0.5)	Random(0.5)	Random(0.5)	Random(0.5)	Random(0.5)
N03	(N04 ⊕ N05)	(N04 ⊕ N05)	N10∧(N04 ⊕ N05)	(N04 ⊕ N05)	(N04 ⊕ N05)
N04	(N03 ⊕ N05)	(N03 ⊕ N05)	(N03 ⊕ N05)	(N03 ⊕ N05)	(N03 ⊕ N05)
N05	(N03 ⊕ N04)	(N03 ⊕ N04)	(N03 ⊕ N04)	(N03 ⊕ N04)	(N03 ⊕ N04)
N06	Random(0.5)	Random(0.5)	Random(0.5)	(N05 ⊕ N08)	(N05 ⊕ N08)
N07	Random(0.5)	Random(0.5)	Random(0.5)	(N07+N08+N09+N10) ≥ 2	¬(N05 ⊕ N08)
N08	(N09 ⊕ N10)	N05∧(N09 ⊕ N10)	N05∧(N09 ⊕ N10)	N03 ⊕ N05	N09 ⊕ N10
N09	(N08 ⊕ N10)	(N08 ⊕ N10)	(N08 ⊕ N10)	(N04+N05+N07+N08)≤ 2	(N08 ⊕ N10)
N10	(N08 ⊕ N09)	(N08 ⊕ N09)	(N08 ⊕ N09)	N06∧(N05 ⊕ N09)	(N08 ⊕ N09)
N11	Random(0.5)	Random(0.5)	Random(0.5)	Random(0.5)	Random(0.5)
N12	Random(0.5)	Random(0.5)	Random(0.5)	Random(0.5)	Random(0.5)

process. Nevertheless, in this paper *(i)* we are analysing Markovian systems of order 1, whose behaviour depends only on the immediately previous step and *(ii)* although TE is not a direct measure of causal effect, the use of short history length and the generation of time series by means of perturbations makes it possible to consider this measure as a way to infer causal effect [5].

4   Robustness of the Method

In this section we illustrate the results concerning the robustness of the results returned by the DCI method with respect to the variance introduced by the homogeneous system generation. We assessed the ranking of the RSs and the transfer entropy values as a function of a sampled distribution of homogeneous system instances and we also compared different ways for generating it. We first describe the test cases used in the analysis; subsequently, we discuss the results in terms of RSs ranking and transfer entropy.

4.1   Test Cases

We chose five paradigmatic systems composed of 12 nodes updated either by means of a boolean function or randomly. The rationale behind the definition of such systems is that, despite their apparent simplicity, they exhibit a non-trivial dynamics, as they are boolean networks, a modelling framework that has obtained remarkable results in simulating real gene regulatory networks [7–9,11]. Nodes update their state in parallel and synchronously. The functional dependences and the update rules of these systems are shown in Table 1. The size of these systems enables us to perform an exhaustive enumeration of all the possible groups, allowing their complete assessment. The systems analysed are initially set to a random initial state and are evolved in time for 500 steps. In order to avoid short attractors and to better observe the relationships among nodes a perturbation is made every 10 steps—nodes are perturbed sequentially, but not in the same order. In *Case 5* we also studied trajectories in which each node in every step has the same probability $p = 1.5\%$ of being perturbed. It is worth remarking that each perturbation is introduced after the system has recovered a stable dynamical situation.

The five instances share a common structure but differ in specific dynamical organizations of some nodes. In *Case 1*, we consider two independent groups of three nodes (namely, *group A* and *group B*), by assigning at each node the *XOR* function of the other two nodes in the group. *Case 2* derives from *Case 1* by introducing in the first node of *group B* a further dependence from the last node of *group A*, hence introducing information transfer from *group A* to *group B*. *Case 3* is a variant of *Case 2* in which a functional dependence of the first node of *group A* from the last node of *group B* is introduced. In *Case 4*, five heterogeneously linked nodes are influenced by *groupA*. The combination of the dynamical rules of the nodes and their initial condition makes the dynamical behaviour of the sixth node always in phase with the triplet. Finally, *Case 5* derives from *Case 1* by adding two nodes whose dynamical behaviour directly depends on nodes of both *group A* and *group B*: these 8 nodes form a group clearly different from the remaining 4 random nodes, as they are interdependent and ruled by deterministic functions.

4.2   Relevant Sets Ranking

The usual way of computing the T_c value consists in generating an instance of an homogeneous system and compute the average of integration and mutual information of its subsets of any size. These values are then used to assess the statistical significance of the DCI of a given subset of the system under observation. The homogeneous system can be generated according to different models and the time series can be of course of different length. We checked the robustness of the results against both criteria.

As for the model for generating the homogeneous system, we considered two possibilities that differ in the distribution probability used. Let $s^1, s^2, \ldots$ be the time series of the system to be analysed, where $s^i = (x_1^i, x_2^i, \ldots, x_n^i)$. Let $\hat{s}^i = (\hat{x}_1^i, \hat{x}_2^i, \ldots, \hat{x}_n^i)$ be the generic state of the homogeneous system time series. In the following, without loosing generality, we suppose that variables are boolean. The two distributions used for generating the homogeneous systems are the following:

i. Compute the frequency f_i of 1s for each variable x_i occurring in the series.[2] Generate a series of states in which the values occur according to the individual frequencies of the variables, i.e. the value of $\hat{x}_j^i$ is sampled from a Bernoulli distribution with parameter f_i.

ii. Compute the global frequency f of 1s occurring in the series. Generate a series of states in which the values occur according to the global frequencies, i.e. the value of $\hat{x}_j^i$ is sampled from a Bernoulli distribution with parameter f.

---

[2] The frequency of 0s is simply $1 - f_i$.

Table 2. Results attained by using model *(i)*. For each of the first five positions in the ranking, the group occurring most frequently is shown, along with its frequency.

Case 1				Case 2				Case 3		
Rank	CRS	p		Rank	CRS	p		Rank	CRS	p
1	8,9,10	0.8		1	3,4,5	0.62		1	9,10	0.92
2	3,4,5	0.8		2	9,10	0.36		2	4,5	0.8
3	9,10	0.5		3	8,9,10	0.2		3	8,9,10	0.62
4	8,9	0.5		4	5,8,9,10	0.22		4	3,9,10	0.38
5	3,4	0.52		5	8,9,10	0.14		5	4,5,8	0.34

Case 4				Case 5		
Rank	CRS	p		Rank	CRS	p
1	4,6	0.6		1	6,7	0.94
2	4,5,6	0.38		2	5,6,7	0.76
3	3,4,5,6	0.28		3	6,7,8	0.74
4	3,4,5,6	0.22		4	4,6,7	0.52
5	4,5,6,8	0.24		5	6,7,9	0.36

Both models capture the idea of preserving some statistical properties of the data to be analysed, while providing a baseline for the estimation of the main quantities of interest. In particular, randomness is introduced with the aim of avoiding structure and make it possible to compute integration and mutual information for a system that does not have relevant sets in its dynamics. The difference between the two models is that the first maintains the individual frequencies for each variable, while the second just assumes an overall frequency of occurrence of the values and it is therefore less accurate than the former.

We compared the rankings produced by using the two models, collecting statistics for 50 homogeneous system independent replicas. We will first present the results from each of the models, assessing the robustness against its inherent variance, and we subsequently compare the results between the two models.

Results for Model *(i)*. Results of model *(i)* are shown in Table 2, where for each of the first fives positions in the ranking, the group occurring most frequently is shown, along with its frequency in that position. The results shown in Table 2 are also confirmed by a statistic on the groups ranked in any of the top five positions: the most frequently occurring groups in each of the top five positions are also those with the highest probability of being ranked among the first five. Results are shown in the appendix (Table 8). The results for *Case 1* are sharp, as the two independent groups of variable are always ranked in the uppermost two positions. The following positions in the rank are occupied by their subsets. Results for *Case 2* are also quite clear: the two dependent groups are ranked in the first positions and their interaction is captured by the detection of groups containing variables from both blocks (e.g. group {N5,N8,N9,N10} ranked in the fourth position for the 20 % of the times). The dynamics of *Case 3* is more complex than the previous cases and this is reflected by the rankings

Table 3. Results attained by using model *(i)* after the application of the sieving algorithm. For each of the first fives positions in the ranking, the group occurring most frequently is shown, along with its frequency.

	Case 5	
Rank	CRS	p
1	6,7	0.94
2	3,4,5	0.74
3	8,9,10	0.78
4	5,7,8	0.52
5	5,6,8	0.52

Table 4. Statistics of rankings over 50 independent draws of model *(i)* homogeneous system. For each test case, the mean and standard deviation of the Spearman rank correlation coefficient is shown.

ρ_s	mean	std. dev.
Case 1	0.991	0.006
Case 2	0.992	0.004
Case 3	0.992	0.004
Case 4	0.984	0.011
Case 5	0.977	0.016

returned. In fact, the two blocks are no longer emerging as candidate RSs, but rather their parts are ranked high. This phenomenon can be explained by observing that pairs of variables are usually way more integrated than triplets. However, the rankings obtained by model *(i)* are still able to capture the essence of system structure. The dependence graph among variables of *Case 4* is rather intricate and so is its dynamics. Nevertheless, the analysis of *Case 4* enlightens some notable groups of interacting variables, such as {N4,N6}, and {N3,N4,N5}. Finally, results of *Case 5* are surprisingly sharp: the two groups of variables (i.e. *group A* and *group B*) are identified, along with the two controlled nodes N6 and N7.

We observe that parts of the same candidate RS are often ranked in the first positions, thus obfuscating the organisation emerging from the analysis. To this aim, the sieving algorithm is indeed applied and a clearer picture of the organisation in terms of RSs is provided. An excerpt of the results of the application of the sieving algorithm is reported in Table 3 (see the appendix for complete data in Table 9). The use of this algorithm makes it possible to clean the picture of the dynamical organisation of the systems and identify its RSs. As an example, let us consider *Case 5*: the three main RSs are robustly ranked in the first positions.

The advantage of using the sieving algorithm might be harmed by the variance introduced by the homogeneous system. To estimate this variance, we compared the rankings by means of the Spearman rank correlation coefficient, which is a special case of the *general correlation coefficient* introduced by Kendall [4]. The coefficient is applied pairwise, considering two rankings r and s:

$$\rho_s = 1 - \frac{6 \sum_{i=1}^{n} (r_i - s_i)^2}{n^2(n^2 - 1)} \tag{8}$$

where r_i and s_i are the rank of element i in the two rankings and n is the number of elements. Note that the coefficient is well defined only in the case of two rankings containing the same elements. By computing ρ_s for every possible pair of

Table 5. Comparison of the rankings between the two models. For each test case, the mean and standard deviation of the Spearman rank correlation coefficient is shown.

ρ_s	mean	std. dev.
Case 1	0.977	0.00003
Case 2	0.986	0.00002
Case 3	0.992	0.00001
Case 4	0.966	0.00018
Case 5	0.968	0.00030

Table 6. Comparison of the rankings obtained with time series of different lengths. For each test case, the mean and standard deviation of the Spearman rank correlation coefficient is shown.

ρ_s	mean	std. dev.
Case 1	0.985	0.009
Case 2	0.991	0.005
Case 3	0.988	0.007
Case 4	0.983	0.010
Case 5	0.972	0.016

the 50 rankings and taking the average, we obtain the results of Table 4. Indeed, the rankings are rather stable. This observation also suggests the possibility of taking the average rank as the main information for the sieving algorithm, so as to dampen fluctuations due to sampling.

Results for Model (ii). The results obtained by applying method (ii) for generating the homogeneous system do not significantly differ from those of method (i), both in terms of relative positions and rank correlation coefficient. For this reason we omit the results.

We conclude by observing that the rankings produced by the two models have negligible differences, as we can observe by the average rank correlation coefficient reported in Table 5. The average is computed over all the possible pairs of rankings. The robustness inter and intra models for generating the homogeneous system guarantees that the application of the DCI method is reliable and stable.

Data Series Length. We also assessed the robustness of the results as a function of both models (i) and (ii) and the length of the data series. We applied the DCI method[3] for data series of length 1, 5, 10, 20, 25 and 30 times the length of the original series. For the sake of brevity, we just report the average rank correlation coefficient computed across all the possible pairs of rankings for all the possible data series lengths (see Table 6). We observe that the rankings are independent of the data series length. This result enables us to assert that a good practice for the application of the DCI method is to generate a homogeneous system data series of the same length as the original one.

5   Transfer Entropy

Once the RSs have been identified, the information flow among them—or at least just correlation—can be quantified by means of TE. The data we have considered in these experiments are time series of the perturbed time evolution

---

[3] As results are not distinguishable for the two models, results just concern model (i).

of a discrete system, therefore the analysis made by means of TE can indeed provide meaningful pieces of knowledge concerning information flow among RSs.

We computed the TE between every pair of RSs. To assess the significance of these values, we compared them with TE computed over the same time series in which the observations are randomly permuted. Such a time series has the same statistical properties of the original one except for the causal relations produced by the boolean update functions. For each time series, we generated 50 random shuffling and computed the TE between the RSs identified. These values were then used to compute a p-value for assessing the significance of the TE values

Table 7. Transfer entropy T between RSs in the five test cases. The values in the table represent $T_{Y \to X}$, where Y is the element in the column and X in the row.

Case 1

TE(col→row)	3,4,5	8,9,10
3,4,5	—	0.001
8,9,10	0.001	—

Case 2

TE(col→row)	3,4,5	8,9,10	9,10	3,4,5,8,9,10	4,5
3,4,5	—	0.005	0.003	0.005	0
8,9,10	0.221	—	0	0.221	0.221
9,10	0.404	0.691	—	0.694	0.379
3,4,5,8,9,10	0	0	0	0	0
4,5	0.117	0.03	0.003	0.118	—

Case 3

TE(col→row)	3,4,5	8,9,10	9,10	3,4,5,8,9,10
3,4,5	—	0.788	0.775	0.788
8,9,10	0.782	—	0	0.782
9,10	0.163	0.838	—	0.853
3,4,5,8,9,10	0	0	0	—

Case 4

TE(col→row)	3,4,5,6	4,5,6,8	4,5,6,10	3,4,6,8,10	4,5,6
3,4,5,6	—	0.212	0.039	0.213	0
4,5,6,8	0.071	—	0.015	0.081	0
4,5,6,10	0.076	0.184	—	0.25	0
3,4,6,8,10	0.078	0.077	0.077	—	0.078
4,5,6	0.096	0.237	0.059	0.309	—

Case 5

TE(col→row)	6,7	8,9,10	3,4,5
6,7	—	0.107	0.116
8,9,10	0.003	—	0.004
3,4,5	0.003	0.004	—

computed on the original time series. Results are shown in Table 7. The TE values corresponding to a p-value ≤ 0.05 are in bold. Each entry of the table contains the TE value computed from the group in the column to the one in the row. We observe that the TE analysis captures the structure of the boolean systems, as in each of the five cases, the significant values of TE correspond to the pairs of groups that actually exchange information. Notably, the actual value of TE might not be sufficiently informative; indeed, we can observe that there are low TE values that turn out to be significant (see, e.g. *Case 4* in Table 7) and, conversely, non negligible values that are instead not significant (see, e.g. *Case 2* in Table 7).

A quantitative comparison among groups of different size can be done by computing a normalised TE. According to [3], the normalised TE (NTE) is defined as:

$$NTE(Y \to X) = \frac{TE(Y \to X) - TE(Y_S \to X))}{h_x} \qquad (9)$$

where $h_x = -\sum_X p(x^{t+1}, x) log\ p(x^{t+1}|x)$ and $TE(Y_S \to)$ is the TE computed on a homogeneous system obtained by randomly shuffling the observations in the data series, as previously described. The values of NTE are computed 50 times, each using a random shuffling of the original data and tables (see Table 10 in the appendix) report mean and standard deviation. It is important to note that these results match quite precisely the functional relations introduced by the boolean functions which impact the dynamics of the system.

6   Conclusion and Future Work

In this work we have assessed the robustness of the DCI method against the homogeneous system. Results show that the method is both robust and reliable. Indeed, the robustness of the method is a requirement for its application in the identification of a plausible and sound hypothesis on the organisation of a dynamical system. It is important to remark that we are interested in the organisation emerging in a system from its dynamics, rather then its static relational structure. As ongoing work, we are devising an improvement over the DCI method that makes it possible to extract information on the hierarchical organisation of a complex systems, thus not just identifying its RSs and the information flow among them, but also their possibly tangled hierarchical organisation.

Appendix

Table 8. Results attained by using model *(i)*. For each group the probability of being ranked in the first five positions is shown.

Case 1		Case 2		Case 3		Case 4		Case 5	
CRS	p	CRS	p	CRS	p	CRS	p	CRS	p
8,9,10	0.98	3,4,5	0.96	8,9,10	1.0	4,5,6	0.98	6,7	1.0
3,4,5	0.8	9,10	0.92	9,10	1.0	4,6	0.98	5,6,7	0.86
3,4	0.7	3,4,5,8	0.74	4,5	0.92	3,4,5,6	0.76	6,7,8	0.84
8,9	0.7	8,9,10	0.5	3,9,10	0.74	3,4,5	0.58	4,6,7	0.58
9,10	0.7	3,4	0.46	4,5,9,10	0.58	4,5,6,8	0.42	5,6,7,8	0.52
4,5	0.18	5,8,9,10	0.42	4,5,8	0.4	4,5	0.36	6,7,9	0.36
4,5,8,9,10	0.16	4,5	0.3	4,5,8,9,10	0.22	5,6	0.22	4,5,6,7	0.32
3,4,5,9	0.16	3,4,8	0.22	3,8,9,10	0.08	3,4,5,6,8	0.18	3,4,5,6,7	0.16
3,4,8,9,10	0.16	3,4,5,6	0.14	9,10,11	0.06	3,4,6	0.14	6,7,8,9	0.12
4,8,9,10	0.16	4,5,8,9,10	0.12			4,5,6,10	0.14	3,4,6,7	0.12
3,4,5,9,10	0.14	4,5,8	0.08			3,8	0.08	6,7,8,9,10	0.08
3,4,5,6	0.12	3,4,5,8,9,10	0.06			3,4,5,6,10	0.08	4,5	0.02
3,4,5,8,9,10	0.04	3,4,8,9,10	0.04			4,6,8	0.06	4,5,6,7,8	0.02
		3,4,5,9,10	0.02			4,5,6,8,10	0.02		
		5,8	0.02						

Table 9. Results attained by using model *(i)* after the application of the sieving algorithm. For each of the first fives positions in the ranking, the group occurring most frequently is shown, along with its frequency.

	Case 1			Case 2			Case 3	
Rank	CRS	p	Rank	CRS	p	Rank	CRS	p
1	8,9,10	0.8	1	3,4,5	0.62	1	9,10	0.92
2	3,4,5	0.8	2	9,10	0.42	2	4,5	0.92
3	3,4,9,10	0.52	3	3,4,8	0.58	3	4,8	0.42
4	4,5,9,10	0.52	4	4,5,8	0.58	4	3,8	0.2
5	3,4,8,9	0.52	5	3,4,7	0.46	5	1,3,5,8	0.34

	Case 4			Case 5	
Rank	CRS	p	Rank	CRS	p
1	4,6	0.6	1	6,7	0.94
2	3,4,5	0.54	2	3,4,5	0.74
3	5,6	0.5	3	8,9,10	0.78
4	3,8	0.34	4	5,7,8	0.52
5	8,10	0.34	5	5,6,8	0.52

Table 10. Normalised transfer entropy T between RSs in the five test cases. The values in the table are the average values and their standard deviation of $NT_{Y \to X}$, where Y is the element in the column and X in the row. Statistics are computed across 50 homogeneous system instances.

Case 1

NTE(col→row)	3,4,5	8,9,10
3,4,5	—	-0.06 ± 0.06
8,9,10	-0.07 ± 0.06	—

Case 2

NTE(col→row)	3,4,5	8,9,10	9,10	3,4,5,8,9,10	4,5
3,4,5	—	-0.13 ± 0.03	-0.07 ± 0.03	-0.47 ± 0.09	0 ± 0
8,9,10	0.48 ± 0.02	—	0 ± 0	0.31 ± 0.04	0.502 ± 0.019
9,10	0.462 ± 0.009	0.844 ± 0.008	—	0.769 ± 0.018	0.44 ± 0.01
3,4,5,8,9,10	0 ± 0	0 ± 0	0 ± 0	—	0 ± 0
4,5	0.525 ± 0.019	0. ± 0.03	-0.07 ± 0.03	0.24 ± 0.05	—

Case 3

NTE(col→row)	3,4,5	8,9,10	9,10	3,4,5,8,9,10
3,4,5	—	0.799 ± 0.01	0.811 ± 0.006	0.66 ± 0.03
8,9,10	0.794 ± 0.013	—	0 ± 0	0.67 ± 0.03
9,10	0.132 ± 0.008	0.884 ± 0.004	—	0.798 ± 0.016
3,4,5,8,9,10	0 ± 0	0 ± 0	0 ± 0	—

Case 4

NTE(col→row)	3,4,5,6	4,5,6,8	4,5,6,10	3,4,6,8,10	4,5,6
3,4,5,6	—	0.465 ± 0.019	0.017 ± 0.016	0.37 ± 0.02	0 ± 0
4,5,6,8	0.146 ± 0.013	—	-0.069 ± 0.019	0.03 ± 0.03	0 ± 0
4,5,6,10	0.107 ± 0.011	0.28 ± 0.012	—	0.35 ± 0.02	0 ± 0
3,4,6,8,10	0.136 ± 0.012	0.088 ± 0.018	0.09 ± 0.02	—	0.077 ± 0.017
4,5,6	0.183 ± 0.01	0.483 ± 0.009	0.084 ± 0.01	0.54 ± 0.02	—

Case 5

NTE(col→row)	6,7	8,9,10	3,4,5
6,7	—	0.26 ± 0.008	0.286 ± 0.01
8,9,10	-0.022 ± 0.018	—	-0.05 ± 0.03
3,4,5	-0.021 ± 0.027	-0.047 ± 0.026	—

References

1. Filisetti, A., Villani, M., Roli, A., Fiorucci, M., Poli, I., Serra, R.: On some properties of information theoretical measures for the study of complex systems. In: Pizzuti, C., Spezzano, G. (eds.) WIVACE 2014. CCIS, vol. 445, pp. 140–150. Springer, Heidelberg (2014)

2. Filisetti, A., Villani, M., Roli, A., Fiorucci, M., Serra, R.: Exploring the organisation of complex systems through the dynamical interactions among their relevant subsets. In: Andrews, P., Caves, L., Doursat, R., Hickinbotham, S., Polack, F., Stepney, S., Taylor, T., Timmis, J. (eds.) Proceedings of the European Conference on Artificial Life 2015 - ECAL2015, pp. 286–293 (2015)
3. Gourévitch, B., Eggermont, J.: Evaluating information transfer between auditory cortical neurons. J. Neurophysiol. **97**, 2533–2543 (2007)
4. Kendall, M.: Rank Correlation Methods. Hafner Publishing Co., New york (1955)
5. Lizier, J., Prokopenko, M.: Differentiating information transfer and causal effect. Eur. Phys. J. B **73**(4), 605–615 (2010)
6. Schreiber, T.: Measuring information transfer. Phys. Rev. Lett. **85**(2), 461–464 (2000)
7. Serra, R., Villani, M., Graudenzi, A., Kauffman, S.: Why a simple model of genetic regulatory networks describes the distribution of avalanches in gene expression data. J. Theor. Biol. **246**(3), 449–460 (2007)
8. Serra, R., Villani, M., Semeria, A.: Genetic network models and statistical properties of gene expression data in knock-out experiments. J. Theor. Biol. **227**, 149–157 (2004)
9. Shmulevich, I., Kauffman, S., Aldana, M.: Eukaryotic cells are dynamically ordered or critical but not chaotic. PNAS **102**(38), 13439–13444 (2005)
10. Tononi, G., McIntosh, A., Russel, D., Edelman, G.: Functional clustering: identifying strongly interactive brain regions in neuroimaging data. Neuroimage **7**, 133–149 (1998)
11. Villani, M., Barbieri, A., Serra, R.: A dynamical model of genetic networks for cell differentiation. PLoS ONE **6**, e17703 (2011)
12. Villani, M., Benedettini, S., Roli, A., Lane, D., Poli, I., Serra, R.: Identifying emergent dynamical structures in network models. In: Bassis, S., Esposito, A., Morabito, F.C. (eds.) Recent Advances of Neural Networks Models and Applications. SIST, vol. 26, pp. 3–13. Springer, Heidelberg (2014)
13. Villani, M., Filisetti, A., Benedettini, S., Roli, A., Lane, D., Serra, R.: The detection of intermediate-level emergent structures and patterns. In: Liò, P., Miglino, O., Nicosia, G., Nolfi, S., Pavone, M. (eds.) Advances in Artificial Life, ECAL 2013, pp. 372–378. The MIT Press, Cambridge (2013). http://mitpress.mit.edu/books/advances-artificial-life-ecal-2013
14. Villani, M., Roli, A., Filisetti, A., Fiorucci, M., Poli, I., Serra, R.: The search for candidate relevant subsets of variables in complex systems. Artif. Life **21**(4), 395–397 (2015)

Dynamically Critical Systems and Power-Law Distributions: Avalanches Revisited

Marina L. Di Stefano[1], Marco Villani[1], Luca La Rocca[1],
Stuart A. Kauffman[2], and Roberto Serra[1(✉)]

[1] Department of Physics, Informatics and Mathematics,
University of Modena and Reggio Emilia, v. Campi 213a, 41125 Modena, Italy
`luce.marina@gmail.com,`
`{marco.villani,luca.larocca,roberto.serra}@unimore.it`
[2] Institute for Systems Biology, Seattle 401 Terry Ave N, Seattle, WA 98109, USA
`stukauffman@gmail.com`

Abstract. In this paper we show that a well-known model of genetic regulatory networks, namely that of Random Boolean Networks (RBNs), allows one to study in depth the relationship between two important properties of complex systems, i.e. dynamical criticality and power-law distributions. The study is based upon an analysis of the response of a RBN to permanent perturbations, that may lead to avalanches of changes in activation levels, whose statistical properties are determined by the same parameter that characterizes the dynamical state of the network (ordered, critical or disordered). Under suitable approximations, in the case of large sparse random networks an analytical expression for the probability density of avalanches of different sizes is proposed, and it is shown that for not-too-small avalanches of critical systems it may be approximated by a power law. In the case of small networks the above-mentioned formula does not maintain its validity, because of the phenomenon of self-interference of avalanches, which is also explored by numerical simulations.

1 Introduction

It has been repeatedly suggested that biological (and perhaps also artificial) evolution should preferentially lead to states that are dynamically critical [1–6]. These states, sometimes said to be "at the edge of chaos", are neither too rigidly ordered nor chaotic; if the system is described by a dynamical system, the claim translates into the statement that evolution should tune the system's parameters, so they should be at (or close to) the separatrices between regions of ordered behavior (where the attractors are, e.g., fixed points or limit cycles) and regions where the attractors are chaotic.

It is also often assumed that the presence of power-law distributions is the hallmark of criticality. Indeed, slightly different (although overlapping) notions of criticality have been used [7]. In this paper we show that a well-known model of genetic regulatory networks, introduced by one of us several years ago [8], i.e. that of Random Boolean Networks (RBNs), can be used to study the relationships between power-law distributions and criticality issues.

© Springer International Publishing Switzerland 2016
F. Rossi et al. (Eds.): WIVACE 2015, CCIS 587, pp. 29–39, 2016.
DOI: 10.1007/978-3-319-32695-5_3

This work is based in part on previous investigations by some of us [9, 10], where it had been shown that RBNs can simulate the statistical properties of the changes induced by single gene knock-out in the expression levels of all the genes of *S. Cerevisiae*. In this paper we are not concerned with the comparison of the model with experimental data, but we rather deepen the analysis of the behavior of the model when subject to small permanent perturbations. The smallest perturbation of this type consists in fixing the value of a single node. Here we will consider perturbations that simulate the knock-out of a randomly chosen gene: among the N nodes of the network, one is chosen at random and its value is fixed to 0. However, as it will be discussed in Sect. 2, RBNs have cyclic attractors, and we perform the perturbation after the network has reached an attractor. It is therefore possible that the candidate node be always 0 in every state of the attractor, but in this case clamping it to 0 would have no effect at all; so we discard that node and we choose another one.

In our studies we then compare the time behavior of the unperturbed ("wild type", briefly WT) network with that of the perturbed one ("knocked-out", KO) that differs from the first by the clamping to 0 of the chosen node (let us call it node R). A node is said to be *affected* if its value in the KO network differs from that in the WT network at least once, after the clamping in root. Since nodes are connected, the perturbation can in principle spread, and it is not limited to node R, or to those nodes that are directly connected to it. The avalanche associated to that particular knock-out is the set of affected genes, and the size of the avalanche is the cardinality of that set (let us call it V). In order to compare results concerning different networks, it is sometimes useful to use the relative size of the avalanche, i.e. the ratio V/N.

One of the most intriguing features of the RBN model is that it allows one to distinguish ordered from disordered (often called "chaotic") regimes on the basis of a single parameter, sometimes called the Derrida parameter λ; as it will be discussed in Sect. 2 this parameter depends upon the choice of the Boolean functions and upon the average number of links per node. Ordered states have $\lambda < 1$, for chaotic states $\lambda > 1$; the value $\lambda = 1$ separates order from chaos, and it is therefore the critical value.

Under the assumptions that the number of incoming links per node A is small ($A \ll N$) and that the overall avalanche is small ($V \ll N$), it can be proven, as it will be shown in Sect. 3, that the distribution of avalanches depends only upon the same Derrida parameter that determines the dynamical regime of the network. The assumptions made here amount to suppose that an avalanche never interferes with itself. Precisely: an affected node B is defined to be the parent of another affected node C if the first deviation of C from the unperturbed value is due to the influence of B. The non-interference condition amounts to assuming that every node C in the avalanche is not affected by any other affected node different from B (neither at a later stage nor at the same time). Therefore, under these assumptions the topology of a spreading avalanche is that of a tree, where each node has a single parent.

The dependency of the avalanche distribution upon λ had already been derived in our previous paper [10], however at that time it was not possible to provide a formula for avalanches of arbitrary size, because a numerical coefficient had to be manually computed. Here, after correcting a missing term, a recursive formula appears to correctly describe the distribution of avalanches up to size 8. It has then been hypothesized that the formula holds for any avalanche size v, an *ansatz* that has been numerically verified on simulated avalanches in networks with 1000 nodes and also theoretically proven.

The correct formula for $p(v) = Pr(V = v)$ is then:

$$p(v) = \frac{v^{v-2}}{(v-1)!} \lambda^{v-1} e^{-\lambda v} \tag{1}$$

Equation 1 is the same as the one that had been previously reported by Ramo [11], but here it is derived in the physically sound "quenched" model, where all the connections and Boolean functions are fixed for each network, without resorting to the "annealed" approximation [12], where connections and Boolean functions are changed at random at each time step, thus losing any possibility of identifying dynamical attractors.

Equation 1 is valid for avalanches of any size and it is not a power law; by inserting the value $\lambda = 1$ we can derive the distribution for avalanches of any size in dynamically critical networks. As it will be shown in Sect. 3, this does not lead to a true power law. However, if we consider fairly large avalanches, for which the Stirling approximation holds (while still being $V \ll N$), it turns out that the distribution indeed approximates a power law with slope $-3/2$.

These results help to clarify the relationship between the concepts of dynamical criticality and those based upon the existence of power-law distributions. In our view, dynamical criticality is a more profound concept, and it may lead (and it often leads) to approximate power-law distributions of interesting quantities.

Due to their modularity, it is sometimes interesting to consider relatively small gene regulatory networks. In these cases the approximation $V \ll N$ may not hold, and an affected node may be subject to the influence of another changed node, so self-interference can take place. We have also numerically explored this phenomenon, by counting the fraction of self-interfering avalanches as a function of the network size. It is shown in Sect. 4 that this fraction can be a substantial one in networks composed by tens or even hundreds of genes. Note that, while real genetic networks usually host thousands of genes, most of those networks that have been described in detail in the literature are relatively small ones; if it were true that their behavior be largely uncoupled from that of the whole network, then the self-interference of avalanches might have relevant biological implications.

Some comments and conclusions are finally drawn in Sect. 5

2 Random Boolean Networks

Here below a synthetic description of the model main properties is presented, referring the reader to [1, 2, 13] for a more detailed account. Several variants of the model have been presented and discussed, but we will restrict our attention here to the "classical" model. A classical RBN is a dynamical system composed of N genes, or nodes, which can take either the value 0 (inactive) or 1 (active). Let $x_i(t) \in \{0,1\}$ be the activation value of node i at time t, and let $X(t) = [x_1(t), x_2(t) \ldots x_N(t)]$ be the vector of activation values of all the genes.

The relationships between genes are represented by directed links and Boolean functions, which model the response of each node to the values of its input nodes. In a classical RBN each node has the same number of incoming connections k_{in}, and its k_{in} input nodes are chosen at random with uniform probability among the remaining $N - 1$ nodes: in such a way the distribution of the outgoing connections per node tends to a Poisson distribution for large N. The Boolean functions can be chosen in different ways: in this paper we will only examine the case where they are chosen at random with uniform probability in a predefined set of allowed transition functions.

In the *quenched* model, both the topology and the Boolean function associated to each node do not change in time. The network dynamics are discrete and synchronous, so fixed points and cycles are the only possible asymptotic states in finite networks (a single RBN can have, and usually has, more than one attractor). The model shows two main dynamical regimes, ordered and disordered, depending upon the degree of connectivity and upon the Boolean functions. Typically, the average cycle length grows as a power of the number of nodes N in the ordered region and diverges exponentially in the disordered region [1]. The dynamically disordered region also shows sensitive dependence upon the initial conditions, not observed in the ordered one.

It should be mentioned that some interesting analytical results have been obtained by the *annealed* approach [12], in which the topology and the Boolean functions associated to the nodes change at each step. Several results for annealed nets hold also for the corresponding ensembles of quenched networks. Although the annealed approximation may be useful for analytical investigations [13], in this work we will always be concerned with quenched RBNs, which are closer to real gene regulatory networks.

A very important aspect concerns how to determine and measure the RBNs' dynamical regime: while several procedures have been proposed, an interesting and well-known method directly measures the spreading of perturbations through the network. This measure involves two parallel runs of the same system, whose initial states differ for only a small fraction of the units. This difference is usually measured by means of the Hamming distance h(t), defined as the number of units that have different activations on the two runs at the same time step (the measure is performed on many different initial condition realizations, so one actually considers the average value $<h(t)>$, but we will omit below the somewhat pedantic brackets). If the two runs converge to the same state, i.e. $h(t) \rightarrow 0$, then the dynamics of the system are robust with respect to small perturbations (a signature of the ordered regime), while if $h(t)$ grows in time (at least initially) then the system is in a disordered state. The critical states are those where $h(t)$ remains initially constant. If a single node is perturbed, the average number of differing nodes at the following time step will be equal to [the probability that a node changes value if one of its input changes] times [the average number of connections per node], a quantity that is sometimes called the Derrida parameter.

In the classical model of RBNs, Boolean functions are often chosen at random among all those with k_{in} values, but a detailed study of tens of actual genetic control circuits [14] has shown that in real biological systems only canalizing functions are found: a function is said to be canalizing if there is at least one value of one of its inputs that uniquely determines the output. Therefore it may be interesting to consider cases where only canalizing functions are allowed. Moreover, if we associate the value 0 to inactivity, a node that is always 0 will never show its presence, so it may be interesting also to consider cases where the null function is excluded [9].

3   Perturbations in RBNs

As discussed in the Introduction, one can compare what happens in the WT RBN and in the KO RBN: at the beginning, a single node (that is, the knocked-out one, also called the root of the perturbation) will differ in the two cases, so the size of the initial avalanche will be 1. If no one of the nodes that receive input from the root changes its value, then the avalanche stops there and it will turn out to be of size 1.

Therefore one can compute p_1, i.e. the probability that an avalanche has size 1, as follows. Let q be the probability that a node chosen at random changes its value if one (and only one) of its inputs changes its value; p_1 is then the probability that all the output nodes of the root do not change, and if there are k outgoing connections, this probability is q^k; therefore, integrating over the outgoing distribution:

$$p_1 = \sum_{k=0}^{N-1} p_{out}(k)q^k \tag{2}$$

where $p_{out}(k)$ is the probability that a node chosen at random has k outgoing connections.

As far as larger avalanches are concerned, we will limit in this section to consider the case of large sparse networks with (on average) a few connections per node; therefore the probability that an output node of the root is also one of its input nodes is negligible. In this case the probability that an avalanche has size 2 equals the probability that only one of the output nodes of the root (i.e. a node at level 1) changes its value, and that the perturbation does not propagate downwards to level 2 (i.e. that nodes which receive connections from the affected node do not change their value). Therefore:

$$p_2 = \sum_{k=0}^{N-1} k p_{out}(k)q^{k-1}(1-q) \sum_{m=0}^{N-k-1} p_{out}(m)q^m \tag{3}$$

By applying the same reasoning, one can continue and compute the probability of avalanches of increasing size. Of course, calculations become more and more cumbersome, as the same size can be achieved in different ways (for example, an avalanche of size 3 may be composed by the root and by two nodes at level 1, none at level 2, or by the root, one node at level 1 and one at level 2).

It is however possible to show that every p_m can be written as a function of the probability generating function $F(q)$ of the outdegree distribution, defined as:

$$F = \sum_{k=0}^{N-1} q^m p_{out}(m) \tag{4}$$

and of its derivatives. Indeed p_1 directly coincides with F (see Eq. 2); noting that $\frac{\partial F}{\partial q} = \sum_{k=0}^{N-1} P_{out}(k)kq^{k-1}$ one can show that p_2 (Eq. 3) can be written as:

$$p_2 = (1-q)F\frac{\partial F}{\partial q} \tag{5}$$

In the same way it can be shown [10] that also the higher order probabilities can be expressed as functions of F and its derivatives.

One can move one step further by taking into account the fact that the outdegree distribution in the (classical) model networks is approximately Poissonian:

$$P_{out}(k) = e^{-A}\frac{A^k}{k!} \tag{6}$$

where $A = <k>$ (note that the average of the number of ingoing connections necessarily equals that of the outgoing connections, so there is no need to specify). In this case Eq. 4 becomes:

$$F = \sum_{k=0}^{N-1} q^k e^{-A}\frac{A^k}{k!} \cong \sum_{k=0}^{\infty} q^k e^{-A}\frac{A^k}{k!} = e^{-1}e^{qA} \tag{7}$$

and therefore, introducing the variable $\lambda = ln(1/F)$ [15]:

$$\lambda = (1-q)A$$
$$F = e^{-\lambda} \tag{8}$$
$$p_n = B_n\lambda^{n-1}e^{-n\lambda}$$

From Eq. 8 one can observe that F, and therefore the avalanche distribution (the coefficient B_n depending only on the graph of perturbation spreading) depends only upon the parameter λ that is the product of two terms, i.e. [probability that a node changes value if one of its input changes]*[average number of connections per node]. Therefore it coincides with the same Derrida parameter defined in Sect. 1.

The computation of the coefficients B_n is lengthy an tedious; it has been explicitly performed [15] up to the avalanche of size 8, and the results are summarized in Table 1.

Table 1. Coefficients of the avalanche distribution

Term	Value	Term	Value	Term	Value	Term	Value
B_1	1	B_3	3/2	B_5	125/24	B_7	16807/720
B_2	1	B_4	16/6	B_6	1296/120	B_8	262144/5040

By looking at the way in which these numbers are generated, the following formula can be suggested [15]:

$$B_n = n^{n-2}/(n-1)! \tag{9}$$

Equation 9 correctly describes the entries of Table 1, and it can be conjectured that it holds for every avalanche size n; taken together with Eq. 8 it leads to Eq. 1 (where the size of the avalanche was denoted by v). In Fig. 1 it is shown that this formula does well approximate the observed distribution of avalanches in simulated RBNs with 1000 nodes (right panel), while the comparison is not satisfactory for small networks (20 nodes, left panel), where self-interference plays a key role; see Sect. 4 for further comments.

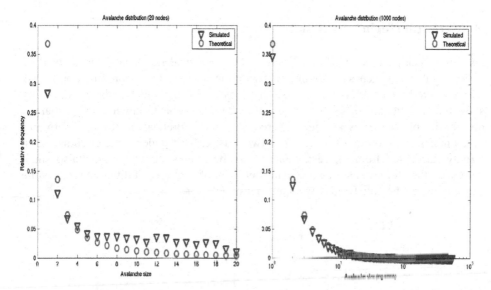

Fig. 1. The theoretical avalanche distribution given by Eq. 1 (red circles) is shown together with the distribution observed in simulations (blue triangles). Every node has exactly 2 inputs, and all the 16 Boolean function are allowed with uniform probability. Left: networks with 20 nodes; right: networks with 1000 nodes (Color figure online)

Let us now come back to the issue of critical systems. The formula for the avalanche distribution (Eq. 1) is valid for avalanches of any size; by inserting the value $\lambda = 1$ we can derive the distribution for avalanches of any size in dynamically critical networks:

$$p(v) = \frac{v^{v-2}}{(v-1)!}e^{-v} \tag{10}$$

It is often stated that power-law distributions are associated with critical states, but Eq. 10 does not describe a power law. However, if we consider fairly large avalanches, such that they still are $v \ll N$, but for which the Stirling approximation holds, i.e. $v! \cong \sqrt{(2\pi v)}(v/e)^v$, we obtain the approximate formula:

$$p(v) \cong (2\pi)^{-\frac{1}{2}}v^{-\frac{3}{2}} \tag{11}$$

that is indeed a power law. This result helps to clarify the relationship between the concepts of dynamical criticality (that, for the reasons given in Sect. 1, appear to be the deeper ones) and those based upon the existence of power-law distributions (that are approximate relationships that often hold at criticality).

It goes without saying that the above remark holds for power-law distributions; close to critical states another type of power-law relationship often holds, which describes the relationship between two different variables (i.e., scaling laws of order parameters as a function of the distance from the critical point).

4   Self-Interfering Avalanches

As it has been observed in Sects. 1 and 3, the theoretical Eq. 1 has been derived by assuming that avalanches do not interfere with themselves. This approximation breaks down when the network is "small", so that it is likely that some avalanches actually show the phenomenon of self-interference. Indeed, it has been shown in Fig. 1 that the distribution of avalanches is largely different from the theoretical one for small networks of 20 nodes. If a portion of a genetic network is, at least under some circumstances, largely decoupled from the rest, then it may be interesting to consider also small networks; therefore we have analyzed the behavior of networks of different sizes, while keeping the connection fixed (two inputs per node).

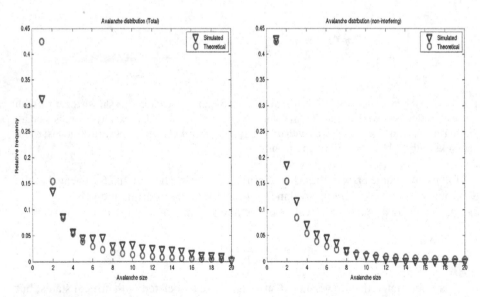

Fig. 2. Distribution of total avalanches (left) and of non-interfering ones (right) for a network with 20 nodes, two connections per node, only canalizing functions allowed

We have developed an algorithm that provides a good approximation to the number of really interfering avalanches, thus separating them from the non-interfering ones. The results obtained for $N = 20$ networks are shown in Fig. 2 and it can be seen that they provide support to our guess that departures from the theoretical formula Eq. 1. are largely due to self-interference. Similar results have been obtained by considering networks of different sizes; as it should be expected, *ceteris paribus* the fraction of interfering avalanches is a monotonous decreasing function of the network size N.

The reported simulations have been performed by considering the case where only canalizing functions are used. As it has been observed, there is a biological reason for that. In this case, the $1-q$ term, i.e. the probability that a node is changed when one of its parent nodes is changed, is no longer ½ (like in the case with all the Boolean functions) but it is rather 3/7. Another biologically interesting case is the one where all the non-canalizing functions and the NULL function are excluded (in this case $1-q = 6/13$). Simulations of avalanche distributions also in this case (not shown here) broadly confirm the above remarks.

5   Conclusions

We have shown here that a very simple formula describes the distribution of non-interfering avalanches of all sizes (provided that they fulfill the non-interference constraint). A similar formula had been obtained by Ramo [11], but by resorting to the annealed approximation. Here the distribution has been explicitly computed for avalanches up to size 8: a recursive formula shows up, so a generalization has been proposed and checked against simulations. It is worth observing that the formula has actually also been proven by the theory of branching process; the interested reader is referred to [15] for details.

The formula allows one to show that approximate power-law distributions can indeed be observed in critical systems for not too small avalanches.

The very interesting phenomenon of avalanche self-interference has been observed and described. It certainly deserves more careful future investigations.

Last but not least, it will be extremely interesting to consider the results of the interaction among different avalanches. This might indeed be the most common case in biology: for example when a chemical is introduced into a cell it is likely to affect more genes at the same time. Interesting effects like the nonlinear dose-response relationships that have been observed might perhaps find at least a partial explanation in the study of the interactions among avalanches.

A final mention of data on real systems is worth, although this paper is not concerned with comparisons with experimental data. It is however interesting to mention that by comparing the experimental data on *S.Cerevisiae* with the theoretical distribution of Eq. 1, using the jackknife method [16, 17], it is possible to locate the λ parameter in the 95 % confidence interval [0.84, 0.93]. Moreover, an analysis of the same data using Bayes factors [18, 19], leads to reject the hypothesis that the network is precisely critical, since the probability that $\lambda = 1$ given the data is smaller than 10^{-3} under a broad range of prior distributions. The interested reader is referred to [15] for further details.

Of course these results are not conclusive, given that we have analyzed a single data set, but they are very interesting. Note also that Kauffman had suggested that living beings might live "in the ordered region, close to the order-chaos border", and this is perfectly compatible with the results of the above analysis. Note also that the claim concerning the advantages of critical states might refer to organisms or colonies, and not necessarily to single cells. The relationship between the dynamics of a single cell and that of an organism (or of an organ) may be far from trivial and we refer the interested reader to [20–23].

Acknowledgments. Useful discussions with Alex Graudenzi, Chiara Damiani and Alessandro Filisetti are gratefully acknowledged.

References

1. Kauffman, S.A.: The Origins of Order. Oxford University Press, New York (1993)
2. Kauffman, S.A.: At Home in the Universe. Oxford University Press, New York (1995)
3. Packard, N.H.: Adaptation toward the edge of chaos. In: Kelso, J.A.S., Mandell, A.J., Shlesinger, M.F. (eds.) Dynamic Patterns in Complex Systems, pp. 293–301. World Scientific, Singapore (1988)
4. Langton, C.G.: Computation at the edge of chaos. Physica D **42**, 12–37 (1990)
5. Langton, C.G.: Life at the edge of chaos. In: Langton, C.G., Taylor, C., Farmer, J.D., Rasmussen, S. (eds.) Artificial Life II, pp. 41–91. Addison-Wesley, Reading MA (1992)
6. Benedettini, S., Villani, M., Roli, A., Serra, R., Manfroni, M., Gagliardi, A., Pinciroli, C., Birattari, M.: Dynamical regimes and learning properties of evolved Boolean networks. Neurocomputing **99**, 111–123 (2013)
7. Bak, P., Tang, C., Wiesenfeld, K.: Self-organized criticality: an explanation of 1/f noise. Phys. Rev. Lett. **59**, 381–384 (1987)
8. Kauffman, S.A.: Metabolic stability and epigenesis in randomly constructed nets. J. Theor. Biol. **22**, 437–467 (1969)
9. Serra, R., Villani, M., Semeria, A.: Genetic network models and statistical properties of gene expression data in knock-out experiments. J. Theor. Biol. **227**, 149–157 (2004)
10. Serra, R., Villani, M., Graudenzi, A., Kauffman, S.A.: Why a simple model of genetic regulatory networks describes the distribution of avalanches in gene expression data. J. Theor. Biol. **249**, 449–460 (2007)
11. Ramo, P., Kesseli, J., Yli-Harja, O.: Perturbation avalanches and criticality in gene regulatory networks. J. Theor. Biol. **242**, 164–170 (2006)
12. Derrida, B., Pomeau, Y.: Random networks of automata: a simple annealed approximation. Europhys. Lett. **1**, 45–49 (1986)
13. Aldana, M., Coppersmith, S., Kadanoff, L.P.: Boolean dynamics with random couplings. In: Kaplan, E., Marsden, J.E., Sreenivasan, K.R. (eds.) Perspectives and Problems in Nonlinear Science, pp. 23–89. Springer, Heidelberg (2003)
14. Harris, S.E., Sawhill, B.K., Wuensche, A., Kauffman, S.A.: A model of transcriptional regulatory networks based on biases in the observed regulation rules. Complexity **7**(4), 23–40 (2001)
15. Di Stefano, M.L.: Perturbazioni in reti booleane casuale. Master thesis, Department of Physics, Informatics and Mathematics, University of Modena and Reggio Emilia (2015)
16. Miller, R.G.: The jackknife—a review. Biometrika **61**, 1–15 (1974)
17. Efron, B.: The Jackknife, the Bootstrap, and Other Resampling Plans. SIAM, Philadelphia (1982)
18. Kass, R.E., Raftery, A.E.: Bayes factors. J. Am. Statist. Assoc. **90**, 773–795 (1995)
19. Robert, C.P.: The Bayesian Choice. Springer, New York (2007)
20. Villani, M., Serra, R., Ingrami, P., Kauffman, S.A.: Coupled random boolean network forming an artificial tissue. In: El Yacoubi, S., Chopard, B., Bandini, S. (eds.) ACRI 2006. LNCS, vol. 4173, pp. 548–556. Springer, Heidelberg (2006)

21. Serra, R., Villani, M., Damiani, C., Graudenzi, A., Colacci, A.: The diffusion of perturbations in a model of coupled random boolean networks. In: Umeo, H., Morishita, S., Nishinari, K., Komatsuzaki, T., Bandini, S. (eds.) ACRI 2008. LNCS, vol. 5191, pp. 315–322. Springer, Heidelberg (2008)
22. Damiani, C., Kauffman, S.A., Serra, R., Villani, M., Colacci, A.: Information transfer among coupled random boolean networks. In: Bandini, S., Manzoni, S., Umeo, H., Vizzari, G. (eds.) ACRI 2010. LNCS, vol. 6350, pp. 1–11. Springer, Heidelberg (2010)
23. Damiani, C., Serra, R., Villani, M., Kauffman, S.A., Colacci, A.: Cell-cell interaction and diversity of emergent behaviours. IET Syst. Biol. 5(2), 137–144 (2011)

Introducing Kimeme, a Novel Platform for Multi-disciplinary Multi-objective Optimization

Giovanni Iacca$^{(\boxtimes)}$ and Ernesto Mininno

Cyber Dyne S.r.l., Via Scipione Crisanzio 119, 70123 Bari, Italy
{giovanni.iacca,ernesto.mininno}@cyberdyne.it

Abstract. Optimization processes are an essential element in many practical applications, such as in engineering, chemistry, logistic, finance, etc. To fill the knowledge gap between practitioners and optimization experts, we developed Kimeme, a new flexible platform for multi-disciplinary optimization. A peculiar feature of Kimeme is that it can be used both for problem and algorithm design. It includes a rich graphical environment, a comprehensive set of post-processing tools, and an open-source library of state-of-the-art single and multi-objective optimization algorithms. In a memetic fashion, algorithms are decomposed into operators, so that users can easily create new optimization methods, just combining built-in operators or creating new ones. Similarly, the optimization process is described according to a data-flow logic, so that it can be seamlessly integrated with external software such as Computed Aided Design & Engineering (CAD/CAE) packages, Matlab, spreadsheets, etc. Finally, Kimeme provides a native distributed computing framework, which allows parallel computations on clusters and heterogeneous LANs. Case studies from industry show that Kimeme can be effortlessly applied to industrial optimization problems, producing robust results also in comparison with other platforms on the market.

Keywords: Multi-disciplinary optimization · Software design · Graphical interface · Distributed computing

1 Introduction

Over the past two decades, Computational Intelligence Optimization (CIO) has become a popular topic among computer scientists and practitioners. The reason for this success is manifold. First of all, industrial and societal problems have become (and still are becoming) ever more challenging, thus requiring robust solvers and algorithms. Despite the advancements in exact methods, e.g. based on classical mathematical optimization, and related tools, such as CPLEX [1], which can efficiently solve many classes of optimization problems, there are still problems whose scale and complexity may hinder their use, including examples of large-scale problems, dynamic problems, problems affected by noise, or black-box

© Springer International Publishing Switzerland 2016
F. Rossi et al. (Eds.): WIVACE 2015, CCIS 587, pp. 40–52, 2016.
DOI: 10.1007/978-3-319-32695-5_4

problems for which a mathematical formulation is not even available [2,3]. In all these cases, where strong guarantees are not really needed (or feasible) but best-effort "good" solutions are enough, heuristic methods (or "meta-heuristics"), such as those offered by CIO (Evolutionary Algorithms, Swarm Intelligence, etc.), are often the only effective solution [4].

Another reason for the success of CIO is that these methods are based on very little assumptions (or none at all) on the problem at hand. These methods are, in fact, black-box, so that virtually any input/output system (i.e. a system where inputs-the problem design variables-are mapped to one or more outputs-the problem metrics to minimize or maximize, or "fitness" in the evolutionary jargon) can be optimized by using them. This property is especially useful for example in many engineering, networking, or logistic problems where an explicit, closed-form mapping between inputs and outputs is not available but is often the output of a domain-specific simulator.

Other reasons for the success of CIO are the fact that meta-heuristics can usually be applied to various problems with minimum coding/engineering effort, and, finally, the fact they are largely available in the literature. A plethora of methods exist nowadays, with different properties of robustness and self-adaptation; still, the family of these algorithms is growing and every year the state-of-the-art in optimization is pushed forward.

Table 1. Some of the most popular optimization software tools. The column "Customizable" indicates if the software allows to implement new optimization algorithms or customize existing ones.

Software	GUI	MOO	Open-source	Customizable	License
HEEDS MDO [5]	Yes	Yes	No	No	Commercial
HyperStudy [6]	Yes	Yes	No	Yes	Commercial
Isight [7]	Yes	Yes	No	No	Commercial
LIONsolver [8]	Yes	Yes	No	No	Commercial
modeFRONTIER [9]	Yes	Yes	No	No	Commercial
MOPS [10]	Yes	Yes	No	No	Commercial
Nexus [11]	Yes	Yes	No	No	Commercial
OpenMDAO [12]	No	Yes	Yes	Yes	Free
Optimus [13]	Yes	Yes	No	No	Commercial
OptiY [14]	Yes	Yes	No	No	Commercial
SmartDO [15]	Yes	Yes	No	No	Commercial
Xtreme [16]	Yes	Yes	No	No	Commercial
μGP [17]	No	Yes	Yes	Yes	Free
Kimeme [18]	Yes	Yes	No	Yes	Commercial

As a consequence of this trend, several software packages have been developed in the last years, which provide off-the-shelf algorithms for solving multi-disciplinary optimization problems. Among these, especially those tools designed for solving multi-objective optimization problems, i.e. problems where multiple conflicting criteria have to be optimized [19], have gained an increasing success and popularity. This is indeed a class of problems that arises in several domains, such as engineering [20,21] or finance [22], and which is therefore extremely relevant in practical problems.

A short list of such tools is reported in Table 1, where GUI and MOO indicate, respectively, if the software has a Graphical User Interface and if it allows for multi-objective optimization[1]. We also indicate if each tool is open-source and if it allows for customization of the optimization algorithms. We should note that we included in the table only multi-disciplinary optimization tools that are specifically based on CIO methods, while we excluded software based on classic techniques for convex optimization, integer linear/non-linear programming and methods addressing combinatorial optimization only. We also excluded those technical software products that are not devoted specifically to optimization but still may include optimization methods, such as Matlab (which provides the Optimization Toolbox), and other CAD/CAE or multi-physics software, as well as multi-disciplinary tools that provide (as an extra feature) one or more, often domain-specific, optimization techniques (see for instance AVL CAMEO [23]).

Clearly, different commercial and open-source free platforms are characterized by different features in terms of usability, openness, modularity, scalability, easiness of interfacing with external packages, etc. Nevertheless, many of these products share similar principles and patterns. For instance, among commercial software, many tools allow the user to describe the optimization problem graphically (with a data-flow or process-flow approach); typically, they provide a variably rich toolbox for post-processing, and a relatively small set of off-the-shelf optimization algorithms (closed-source, either legacy or from the literature) with limited possibility for parameter tuning or other algorithm modifications. Instead, open-source tools generally lack complete graphical interfaces or advanced post-processing features, thus resulting more difficult to use, at least for practitioners interested in ready-to-use tools; also, they require in general some knowledge about optimization and programming, for instance for writing scripts or markup language files to interface with third-party software; however, due to their openness, these platforms allow a higher level of flexibility (e.g. expert users can develop their own algorithms or modify existing ones relatively easily).

From this short summary we can note that, as a general trend, commercial and open-source optimization tools are rather far apart in terms of usability and flexibility (but this distinction, arguably, affects all technical software): on one hand, commercial tools prefer usability over flexibility, as they typically address the needs of users who are not necessarily specialized in programming and optimization (such as mechanical engineers, designers, logistic experts, and other corporate technical figures for which optimization is simply a tool).

---

[1] A more complete list is available at: http://en.wikipedia.org/wiki/List_of_optimization_software.

On the other hand, open-source software usually look at academic users, such as mathematicians and optimization scholars who are, in general, more interested in designing novel algorithms and testing them on benchmark functions, rather than applying existing algorithms to specific real-world problems. At the time being, it is hard to find tools that bridge this gap between these two worlds, providing at the same time the robustness and ease of proprietary software and the flexibility of open-source platforms.

Motivated by this idea, we developed a novel commercial multi-disciplinary optimization platform, Kimeme. Similar to other commercial platforms, Kimeme provides, off-the-shelf, a rich graphical interface including the possibility of describing the optimization problem graphically, a large set of post-processing tools, such as plots and statistic analysis, and several CIO methods for both single and multi-objective optimization. Moreover, it includes a native distributed computing system, which allows for a seamless parallelization of the solution evaluations on a local network. However, differently from most commercial software, Kimeme adds the possibility of modifying the code of the existing algorithms, reusing it to design novel algorithms, or even integrating algorithms designed from scratch. In this way, Kimeme tries to bridge the gap between academy and industry, in such a way that ever more powerful algorithms proposed in the specialized literature on optimization can be easily implemented and made immediately available to practitioners who may be willing to apply them to complex real-world problems.

In this paper we present the main features of Kimeme, its architecture and the way algorithms are implemented. First, we describe in Sects. 2.1 and 2.2, respectively, the problem and algorithm design environments, while in Sect. 2.3 we introduce briefly some of the post-processing tools available in the platform. Section 2.4 presents the distributed computing framework integrated in Kimeme, which leverages multiple computers to solve computationally expensive problems. Section 3 presents two case studies to exemplify some possible applications of Kimeme. Finally, in Sect. 4 we give the conclusions of this work.

2   Architecture of Kimeme

The main feature of Kimeme is a rich GUI that assists the user in all the steps of the optimization process, from the problem definition to the results post-processing. In this section, we describe the main components of the Kimeme GUI, namely the problem and algorithm design and the post-processing tools. We also introduce another important feature of the platform, that is its native distributed computing infrastructure.

2.1   Problem Design

In Kimeme, every problem is internally represented as an *evaluation tree*, i.e. an execution tree where each node concurs (in a data-flow logic) to the evaluation of a single solution to the problem at hand. As shown in Fig. 1.a, the tree can

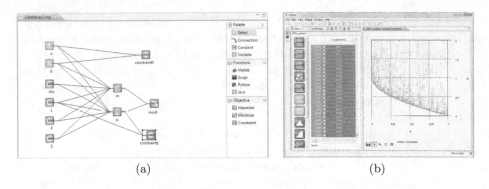

(a)   (b)

Fig. 1. Some screenshots of Kimeme: (a) project design view; (b) solution table view, including a 2D scatter plot (see also Sect. 2.3).

be created visually, simply adding nodes from a palette. The palette currently includes several nodes for defining design variables, constraints, objectives, constants, as well interface nodes to external software, such as Matlab, Python or Java code, Bash or DOS scripts, etc.

2.2   Algorithm Design

A key element of Kimeme is the possibility, for the user, to modify existing algorithms provided off-the-shelf by the platform, or implementing new ones. A set of state-of-the-art optimization algorithms is available, both for single-objective problems (such as Differential Evolution (DE) [24], Evolution Strategies (ES) [25], Self-adapting Differential Evolution (jDE) [26], and Nelder-Mead Simplex [27]) and multi-objective problems, including Non-dominated Sorting Genetic Algorithm 2 (NSGA-2) [28], Multi-objective Particle Swarm Optimization (MOPSO) [29], Strength Pareto Evolutionary Algorithm 2 (SPEA2) [30], Archived Multi-objective Simulated Annealing (AMOSA) [31] and two custom variants of Multi-objective DE (MODE) and ES (MOES).

Kimeme was designed with code re-usability in mind: algorithms are, in fact, fully open-source and structured in such a way that every single component of an algorithm can be reused in another algorithm. Inspired by the modern wave of Memetic Computing [32–36], algorithms are structured in self-contained elements called *operators* (in the memetic jargon, a generalization of a "meme"), each one performing simple operations on the problem solutions. As shown in Fig. 2, the generic algorithm structure in Kimeme consists of a Design of Experiment (DoE), which generates the initial candidate solutions for the problem at hand (i.e., the initial "population"), followed by a sequence of *Start Operators*, *Step Operators* and *Stop Operators*. The first ones are executed, only once, at the beginning of the optimization, to perform various initializations needed for the execution of the algorithm. Step Operators represent the core of the algorithm iterations and are executed, repeatedly, until one or more stop condition is met. Finally, the Stop Operators implement various stop conditions and check at the end of each iteration if any of those conditions is met.

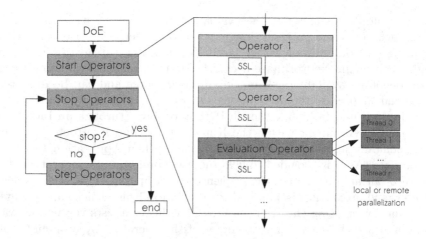

Fig. 2. Generic algorithm flow-chart in Kimeme.

Fig. 3. Structure of the Solution Set List.

The way information is passed among operators is by means of a *Solution Set List* (SSL), i.e. a structured list of candidate solutions (the latter being in turn a structure containing lists of design variables, constraint and fitness values, as well as extra properties of each solution), as depicted in Fig. 3. In this framework, each operator accepts as input a Solution Set List (generated by another operator), and returns as output a new Solution Set List, modified according to its internal operator logics. This structure has two main advantages: on one hand, the use of a structured list of solutions allows an immediate implementation of various modern schemes where multiple populations (composing a "meta-population") co-evolve in parallel, for instance according to an island model, and eventually exchange solutions through some sort of migration mechanisms [37]. Also, this structure is especially useful in multi-objective optimization, where it can be needed to rank the solutions according to their non-domination, thus maintaining different solutions sets, one for each rank.

Kimeme comes with a rich library containing several operators for DoE, crossover, mutation, selection, constraint handling, solution niching and ranking, various utility functions for manipulating the Solution Set List (e.g. merging, cutting, splitting, copying, etc.), different stop conditions, and a number of algorithm-specific operations. Each operator defines one or more specific parameters and additional properties that are needed by the algorithm, and that can also be passed from one operator to another. A special operator, called *Evaluation Operator*, performs the actual evaluation of the incoming SSL, executing multiple instances of the evaluation tree (one for each solution in the SSL) defined in the problem design. Also, this operator is responsible for distributing the computation either locally, on multiple threads, or remotely, through the Kimeme Network (see Sect. 2.4). Apart from the Evaluation Operator, all operators available in Kimeme are open-source and can be dissected, modified and adapted to the users' needs. Operators can be easily implemented in an object-oriented fashion, either in Python or Java (through an Integrated Development Environment within the Kimeme GUI) and interpreted or compiled dynamically at runtime. A complete Java/Python Application Programming Interface (API) is provided to help the users programming operators. The structure of an algorithm (i.e. the sequence of its operators, as well as their parameters and properties) is simply defined by an xml file, which can be edited either manually or using the Kimeme GUI. The latter assists the user in various operations such as changing the order of the operators by drag-and-drop, removing or adding new operators, defining their parameters and properties etc. A checker is also provided to automatically verify that there are no errors in the operators' code or inconsistencies in the logics of the algorithm.

2.3   Post Processing

To analyze the results of the optimization process, Kimeme provides a rich set of plots, statistical analyses and post-processing tools. Some examples of such tools are shown in Fig. 4, including for instance various scatter plots 2D and 3D, and other multi-dimensional visualization plots such as matrix and parallel plots. The typical post-processing use case involves the selection of one or more solutions from the main solution tables (for example those belonging to the Pareto front), and the choice of the desired plots, see Fig. 1.b for an example.

In general, multiple plots can be associated to a single table, providing different levels of information about the solutions generated by the optimization process (for instance their distribution in the search space, the correlation among variables, etc.). The user can easily interact with the plots, for instance zooming in/out, rotating the 3D views, checking a solution and visualizing its details, etc. The plot graphical details (line width, colors, markers, etc.) can also be edited, independently for each plot. Additionally, solutions data and plots can be exported to ASCII, Excel and Matlab files, and various image formats.

2.4   Kimeme Network

We conclude this section with a brief description of the distributed computing feature available in Kimeme, called *Kimeme Network*. Distributed computing is particularly relevant in expensive optimization, for instance in many engineering design problems where each solution evaluation corresponds to a computationally heavy multi-physics simulation. In such scenarios, running an optimization algorithm (which typically requires hundreds of thousands evaluations) on a single computer might introduce a bottleneck in the business processes. To cut design

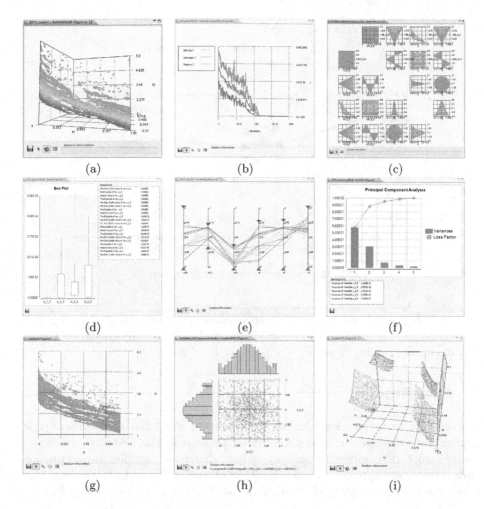

Fig. 4. Some of the post-processing plots available in Kimeme: (a) 3D bubble plot; (b) generation plot; (c) matrix plot; (d) box plot; (e) parallel plot; (f) PCA plot; (g) 2D scatter plot; (h) 2D scatter plot with probability density function; (i) 3D scatter plot.

cost and time, exploiting the inherent parallel nature of most meta-heuristics, distributed computing is therefore needed.

The main architecture of the Kimeme Network is shown in Fig. 5: a central Java daemon, called *Dispatcher*, receives requests for computation (i.e., individual solution evaluations) from one or more optimization processes instantiated by (different instances of) the Kimeme GUI. The Dispatcher then distributes the computations, according to various scheduling rules, on a list of available computing nodes (that can be modified at runtime), which run another Java daemon called *Worker*. Each Worker defines how many CPUs it has available and how many threads it can launch. Both daemons are platform-independent,

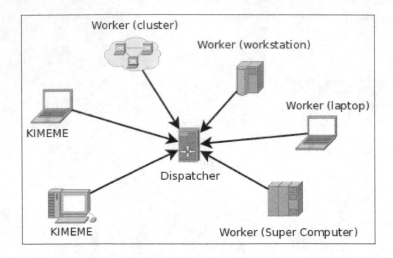

Fig. 5. Kimeme network architecture.

so that it is possible to set up a heterogeneous distributed computing network infrastructure, for instance including Windows/Linux desktops or laptops, next to a cluster and a dedicated high-computing machine.

3   Case Studies

We describe here two case studies of application of Kimeme, to exemplify its applicability to engineering and industrial problems. The first case study is a classic structural engineering design problem, shown here for illustration purpose only. The second application is a complex multi-objective metallurgical problem related to the optimization of productivity and CO_2 emission of a blast furnace.

3.1   Optimal Design of a Cantilever

This problem consists in designing an aluminum cantilever having a fixed length (l) and a fixed force (F) applied on the free edge, as shown in Fig. 6.a. The goal of the design is to keep the structure light-weighted and rigid or, formally speaking, minimizing both the mass (m) and the deflection (w). Decision variables are the outer (a) and the inner (b) edge lengths. The design has to comply with some dimensional and functional constraints, related to the feasibility of the structure (the outer edge length must be grater than the inner one) and to the deflection-length ratio, which must be less than 10 %. A complete formulation of the problem is available in [38].

Figure 6.b shows the Pareto front generated by the MODE algorithm available in Kimeme. We can observe a nice spread of solutions over the Pareto front, which is covered entirely, and quite evenly.

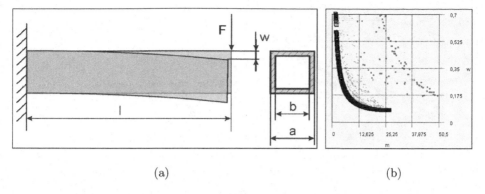

(a)   (b)

Fig. 6. Cantilever design problem: (a) problem description and (b) solutions obtained in Kimeme using MODE with the standard parameter setting (population size 100, 250 generations): initial generation (red), last generation (green), Pareto front at the last generation (black dots) (Color figure online).

3.2 Blast Furnace Productivity/CO2 Emission Optimization

This problem consists in minimizing the CO2 emission of an industrial iron blast furnace while simultaneously maximizing its productivity, with a constraint on the Silicon content of the hot metal produced by the furnace. This case study was developed at the Department of Metallurgical and Materials Engineering at the Indian Institute of Technology, and its results have been originally published in a recent paper by Jha et al. [39]. In this work, the authors performed a systematic comparison between various algorithms available in Kimeme and mode-FRONTIER [9]. They also compared these algorithms with a custom-developed surrogate-assisted algorithm called Evolutionary Neural Network (EvoNN) [40]. In the study, different levels of Silicon were considered. Here, we report the preliminary results of a smaller subset of algorithms (without tuning, unpublished data) obtained on only two configurations (low and medium, respectively with 0.40–0.55 % and 0.55–0.70 % Si). We refer the interested reader to the work by Jha et al. [39] for further details about the problem formulation and for complete optimization results with tuned algorithms.

Figure 7 shows the Pareto fronts found by five algorithms from Kimeme (NSGA-2, MOPSO, MODE, MOES and SPEA2), one from modeFRONTIER (NSGA-2), and EvoNN, in low and medium Silicon level. It should be noted that, among all the methods shown in the figure, only EvoNN is surrogate-assisted, therefore a fair direct comparison between this method and the other methods is not possible. It can be observed (as also reported in [39]) that all the methods implemented in Kimeme guarantee a rather good solution spread over the Pareto front. Compared with the implementation of NSGA-2 in modeFRONTIER, all the algorithms in Kimeme perform quite well, especially MOPSO and MODE that consistently show good results at all Silicon levels.

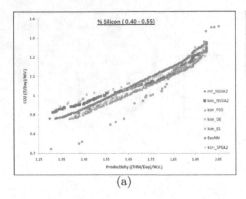 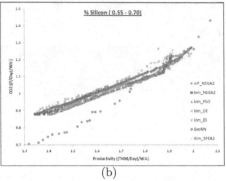

(a)   (b)

Fig. 7. Pareto fronts obtained on the furnace optimization problem described in [39]: (a) low and (b) medium Silicon level. mF_NSGA2 indicates the NSGA-2 implementation in modeFRONTIER, while kim_NSGA2, kim_PSO, kim_DE, kim_ES, kim_SPEA2 indicate, respectively, the open implementation of NSGA-2, MOPSO, MODE, MOES and SPEA2 in Kimeme. Lastly, EvoNN indicates the surrogate-assisted method proposed in [40]. Courtesy of Nirupam Chakraborti.

4   Conclusions

In this paper we have introduced Kimeme, an innovative platform for designing novel optimization algorithms and applying them to real-world problems. The core idea of Kimeme is to bridge the gap between computer scientists and practitioners, so to foster a mutually beneficial transfer of knowledge between academy and industry. The platform offers a configurable environment for both designing algorithms and solving optimization problems, that can be effortlessly tailored to specific applications from various domains. In addition to that, Kimeme provides a very flexible, scalable and easy-to-use distributed computing infrastructure, that can be used to speed up the optimization process. Moreover, the set of state-of-the-art algorithms provided in Kimeme is broad enough to obtain good results on optimization problems from different domains: the case studies presented here showed that Kimeme could be easily applied to problems from mechanical design and metallurgy, but the platform is naturally applicable to other domains. Further developments are planned in the years to come in order to enrich the platform with new features, optimization algorithms and post-processing capabilities.

References

1. IBM: CPLEX. http://www-01.ibm.com/software/commerce/optimization/cplex-optimizer/
2. Tenne, Y., Goh, C.K.: Computational Intelligence in Optimization. Springer, Heidelberg (2010)
3. Koziel, S., Yang, X.S.: Computational Optimization, Methods and Algorithms, vol. 356. Springer, Heidelberg (2011)

4. Zelinka, I., Snasel, V., Abraham, A.: Handbook of Optimization: From Classical to Modern Approach, vol. 38. Springer, Heidelberg (2012)
5. Red Cedar Technology: HEEDS® MDO. http://www.redcedartech.com
6. Altair: HyperStudy©. http://www.altairhyperworks.com
7. Dassault Systèmes: Isight©. http://www.3ds.com
8. LIONlab: LIONsolver. http://lionoso.com/
9. ESTECO: modeFRONTIER®. http://www.esteco.com/modefrontier
10. German Aerospace Center, Institute of System Dynamics and Control, Air-craftSystems Dynamics: MOPS. http://www.dlr.de/rm/en/desktopdefault.aspx/tabid-3842/6343_read-9099/
11. iChrome: Nexus©. http://ichrome.com/solutions/nexus
12. NASA Glenn Research Center: OpenMDAO. http://openmdao.org/
13. Wilde Analysis Ltd.: Optimus®. http://wildeanalysis.co.uk/fea/software/optimus
14. OptiY GmbH: OptiY©. http://www.optiy.eu/
15. FEA-Opt Technology: SmartDO©. http://www.smartdo.co/
16. Optimal Computing: Xtreme©. http://www.optimalcomputing.be/
17. Sanchez, E., Schillaci, M., Squillero, G.: Evolutionary Optimization: The μGP Toolkit, 1st edn. Springer Publishing Company Inc., Berlin (2011)
18. Cyber Dyne Srl: Kimeme. http://cyberdynesoft.it/
19. Deb, K.: Multi-objective optimization. In: Burke, E.K., Kendall, C. (eds.) Search Methodologies, pp. 403–449. Springer, Heidelberg (2014)
20. Köppen, M., Schaefer, G., Abraham, A.: Intelligent Computational Optimization in Engineering, pp. 300–331. Springer, Heidelberg (2011)
21. Yang, X.S., Koziel, S.: Computational Optimization and Applications in Engineering and Industry, vol. 359. Springer, Heidelberg (2011)
22. Chen, S.H., Wang, P.P.: Computational Intelligence in Economics and Finance. Springer, Heidelberg (2004)
23. AVL: AVL^{TM} CAMEO. https://www.avl.com/cameo
24. Storn, R., Price, K.: Differential evolution - a simple and efficient adaptive scheme for global optimization over continuous spaces. J. Global Optim. **11**(TR–95–012), 341–359 (1997)
25. Beyer, H.G.: The Theory of Evolution Strategies. Springer, Heidelberg (2001)
26. Brest, J., Greiner, S., Bošković, B., Mernik, M., Žumer, V.: Self-adapting control parameters in differential evolution: a comparative study on numerical benchmark problems. IEEE Trans. Evol. Comput. **10**(6), 646–657 (2006)
27. Nelder, A., Mead, R.: A simplex method for function optimization. Comput. J. **7**, 308–313 (1965)
28. Deb, K., Pratap, A., Agarwal, S., Meyarivan, T.: A fast and elitist multiobjective genetic algorithm: NSGA-II. IEEE Trans. Evol. Comput. **6**(2), 182–197 (2002)
29. Coello Coello, C.A., Lechuga, M.: MOPSO: a proposal for multiple objective particle swarm optimization. In: Proceedings of the 2002 Congress on Evolutionary Computation, 2002. CEC 2002, vol. 2, pp. 1051–1056 (2002)
30. Zitzler, E., Laumanns, M., Thiele, L.: SPEA2: Improving the Strength Pareto Evolutionary Algorithm (2001)
31. Bandyopadhyay, S., Saha, S., Maulik, U., Deb, K.: A simulated annealing-based multiobjective optimization algorithm: AMOSA. IEEE Trans. Evol. Comput. **12**(3), 269–283 (2008)
32. Iacca, G., Neri, F., Mininno, E., Ong, Y.S., Lim, M.H.: Ockham's razor in memetic computing: three stage optimal memetic exploration. Inf. Sci. **188**, 17–43 (2012)
33. Caraffini, F., Neri, F., Iacca, G., Mol, A.: Parallel memetic structures. Inf. Sci. **227**, 60–82 (2013)

34. Iacca, G., Caraffini, F., Neri, F.: Memory-saving memetic computing for path-following mobile robots. Appl. Soft Comput. **13**(4), 2003–2016 (2013)
35. Neri, F., Cotta, C., Moscato, P.: Handbook of Memetic Algorithms. Studies in Computational Intelligence, vol. 379. Springer, Heidelberg (2011)
36. Caraffini, F., Iacca, G., Neri, F., Mininno, E.: The importance of being structured: a comparative study on multi stage memetic approaches. In: 2012 12th UK Workshop on Computational Intelligence (UKCI), pp. 1–8. IEEE (2012)
37. Mühlenbein, H.: Parallel genetic algorithms, population genetics and combinatorial optimization. In: Becker, J.D., Eisele, I., Mündemann, F.W. (eds.) Parallelism, Learning, Evolution. LNCS, vol. 565, pp. 398–406. Springer, Heidelberg (1991)
38. Cyber Dyne Srl: Kimeme Quick Guide
39. Jha, R., Sen, P.K., Chakraborti, N.: Multi-objective genetic algorithms and genetic programming models for minimizing input carbon rates in a blast furnace compared with a conventional analytic approach. Steel Res. Int. **85**(2), 219–232 (2014)
40. Pettersson, F., Chakraborti, N., Saxén, H.: A genetic algorithms based multi-objective neural net applied to noisy blast furnace data. Appl. Soft Comput. **7**(1), 387–397 (2007)

Optimizing Feed-Forward Neural Network Topology by Multi-objective Evolutionary Algorithms: A Comparative Study on Biomedical Datasets

Vitoantonio Bevilacqua[1]($\boxtimes$), Fabio Cassano[1], Ernesto Mininno[2],
and Giovanni Iacca[2]

[1] Dipartimento di Ingegneria Elettrica e dell'Informazione, Politecnico di Bari,
via Orabona 4, 70125 Bari, Italy
vitoantonio.bevilacqua@poliba.it
[2] Cyber Dyne S.r.l., Via Scipione Crisanzio 119, 70123 Bari, Italy

Abstract. The design of robust classifiers, for instance Artificial Neural Networks (ANNs), is a critical aspect in all complex pattern recognition or classification tasks. Poor design choices may undermine the ability of the system to correctly classify the data samples. In this context, evolutionary techniques have proven particularly successful in exploring the complex state-space underlying the design of ANNs. Here, we report an extensive comparative study on the application of several modern Multi-Objective Evolutionary Algorithms to the design and training of an ANN for the classification of samples from two different biomedical datasets. Numerical results show that different algorithms have different strengths and weaknesses, leading to ANNs characterized by different levels of classification accuracy and network complexity.

Keywords: Artificial Neural Networks · Multi-Objective Evolutionary Algorithms · Akaike Information Criterion

1 Introduction

Over the past two decades, the amount of data produced yearly in all human applications has reached an unprecedented level. As the quantity of data generated in the most complex engineering, networking, and financial systems, is for obvious reasons, impossible to analyze manually, the need arises for expert systems capable of analyzing the data automatically, for instance for the purpose of classification, pattern recognition, and feature extraction. This need is bringing Machine Learning (ML) to a new level, and novel approaches are being presented in the literature, specifically tailored for problems of increasing complexity.

Interestingly, many modern ML classification techniques are now based on Artificial Neural Networks (ANNs). In fact, despite being one of the oldest computational tools known in ML, ANNs are still today among the most effective

© Springer International Publishing Switzerland 2016
F. Rossi et al. (Eds.): WIVACE 2015, CCIS 587, pp. 53–64, 2016.
DOI: 10.1007/978-3-319-32695-5_5

techniques available to solve most kinds of classification problems. On the other hand, while ANNs are powerful learners per se, their performance on specific datasets can be severely affected by several factors. First of all, since neural networks need to learn from examples, typically the training and validation sets must be quite large and should contain balanced class examples. The second problem is feature selection, i.e. the choice of which features should be used as input to the classifier. More features do not lead necessarily to a better classification accuracy, however feature selection can be especially hard in some cases.

One of the most prominent areas of application of ANNs is nowadays health care and health improvement, see e.g. [1,2]. For example, computerized medical imaging systems are constantly improving their ability to extract numerical features from biological data, features that can be used in expert systems (based on ANNs) to assist diagnosis and therapy. Typically, in order to train a robust expert system and obtain a high classification accuracy, one needs a large set of labeled samples. However, in most cases, data acquisition and labeling is expensive (due to the cost of the medical tests, and to the need for a human expert to label the training samples) and the expert system must be trained on a relatively small dataset. Therefore, a third major challenge is to reach a high accuracy with a limited number of labeled samples.

Finally, on top of all the above mentioned problems, there is the choice of the ANN topology, namely the number of layers, the number of nodes per layer, and the activation function used in each node. While simple rules of thumb exist for such a choice, there is no way to predict which configuration of the network is the best to use in each case and one often has to rely on manual trial-and-error. However, the training of each different network configuration is a time consuming process and trial-and-error is obviously prone to sub-optimal results.

Thanks to the ever-increasing availability of computing power, a viable alternative for solving these problems is now the use of automatic techniques that explore the entire space of solutions defined by the ANN topologies, while performing the training of each network on multiple shuffled versions of the dataset at hand. One such example is presented in [3], where Multi-Objective Genetic Algorithm (MOGA) [4] is used to find the optimal ANN topology (i.e., the optimal number of hidden layers and the number of nodes for each hidden layer) which leads to the best classification of the samples from the Wisconsin Breast Cancer Dataset (WBCD) [5].

In this paper, we follow up on [3] by performing an extensive comparison of a whole set of Multi-Objective Evolutionary Algorithms (MOEAs) on two different datasets, namely the aforementioned WBCD and the Hepatitis Dataset (HD)[5]. First, we try to obtain on each dataset the best possible accuracy, by minimizing at the same time the validation and test error. In a second set of experiments, we try to identify the best trade-off between accuracy and network complexity: in this latter case, the optimization criteria are the minimization of (1) the validation error, and (2) a measure of the network complexity, i.e. the Akaike Information Criterion [6] (rather than an explicit minimization of the number of hidden layers and hidden nodes per layer).

The rest of this paper is organized as follows. The next section briefly summarizes the related work on the use of MOEAs for automatic design of classifiers. Section 3 describes the MOEA-based method used in the study, while numerical results are reported in Sect. 4. Finally, Sect. 5 concludes this work.

2   Related Work

MOEAs are bio-inspired multi-objective optimization techniques that have been successfully used in several applications domains, such as engineering design [7,8] and combinatorial optimization [9]. Recently, MOEAs have also been used in real-time applications, as shown in [10], and biomedical applications [11,12].

In the ML domain, there are several examples of application of (either single-objective or multi-objective) Evolutionary Algorithms to the optimization of neural network topology (see e.g. [13]), or for training ANNs [14]. Another example is given by [15], where an improved version of classic Genetic Algorithms (GA) is introduced, specifically designed to optimize the structure of a neural network. In the aforementioned work [4], Multi-Objective Genetic Algorithm (MOGA) has been used to find the best topology in order to improve the neural network accuracy on the WBCD dataset. A similar technique was also proposed in a more recent study [16]. Other biomedical applications of optimized neural networks are also presented in [17–19].

3   Proposed Approach

As mentioned earlier, the main idea of this study is to formulate the problem of the definition of an optimal ANN for a specific dataset in a multi-objective fashion. For example, in order to maximize the accuracy, a Multi-Objective Evolutionary Algorithm can be used to minimize the validation error while minimizing the test error. However, depending on the requirements one could also include different optimization criteria in the problem formulation, such as a measure of complexity of the classifier ANN. We will show this in the next section.

In a nutshell, the proposed multi-objective approach consists of two nested loops, as depicted in Fig. 1: (1) an outer loop, where populations of candidate ANNs are generated and optimized by a MOEA; (2) an inner loop, where each candidate ANN is trained, validated and tested.

More specifically, in the first step, the MOEA defines for each candidate ANN the number of hidden layers, the number of nodes per layer, and (optionally) the activation function that must be used in each layer. This information is then used to create neural networks that are structured as follows: (1) an input layer made of as many nodes as the number of the features of the dataset, with no bias; (2) a variable number of hidden layers, each of which is made of a variable number of nodes (as determined by the MOEA), with bias; (3) an output layer, with a single node (the classification value), with no bias. For every layer, it is possible to select the activation function a priori, or have the MOEA select it: as shown in the next section, in our experiments we tested both options.

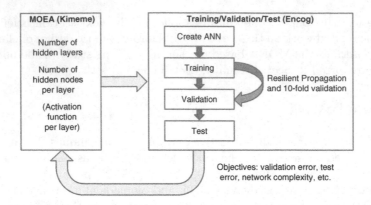

Fig. 1. Conceptual scheme of the MOEA-based approach.

In the second step, each ANN so generated, is then trained, validated and tested on the dataset. This procedure is performed as follows. First, the original dataset is shuffled and partitioned into three sets (in our experiments, 60 %, 20 %, and 20 % of the entire dataset, respectively). Then, the first set is used to train and validate the neural network topology, by means of a 10-fold cross validation. The training method we use in our experiments is the Resilient Propagation [20], with stop condition on a training error threshold. The second and third sets are finally used to calculate, respectively, the validation error, the test error, and the confusion matrix. This information (or, if needed, a metric of network complexity) is then fed back to the outer loop, and used to calculate the fitness functions to be optimized by the MOEA.

In all our experiments (see the next section for further details), we used some of the state-of-the-art MOEAs available in Kimeme, a multi-disciplinary optimization platform introduced in [21,22]. The reason for using Kimeme was manifold: first of all, Kimeme provides a rich set of state-of-the-art single and multi-objective optimization algorithms, as well as an extensive post-processing toolkit. Secondly, Kimeme can be easily coupled with external software and pieces of code (such as Java or Python classes, or Matlab scripts). Importantly, Kimeme also integrates a distributed computing framework, which allowed us to easily run massively parallel calculations. As for the ANN implementation, we used the open-source Java library Encog [23], which is characterized by a great flexibility in the definition of neural networks and training algorithms.

4   Numerical Results

In the following, first we describe our experimental setup (datasets and multi-objective algorithms), then we analyze the numerical results obtained in the different experiments with the approach described in the previous section. We finally report a brief analysis of the execution times of our experiments.

4.1   Datasets

In our experimental study, we consider two biomedical datasets:

- The Wisconsin Breast Cancer Dataset (WBCD) [5]. The WBCD is composed of 699 labeled samples, each defined by 9 biomedical features, namely: clump thickness, uniformity of cell size, uniformity of cell shape, marginal adhesion, single epithelial cell size, bare nuclei, bland chromatin, normal nucleoli, and mitoses. The dataset contains 16 samples with one or more missing values, which we omit from our analysis.
- The Hepatitis Dataset (HD) [5]. The dataset is composed of 155 labeled samples, each defined by 19, namely: age, sex, steroid, antivirals, fatigue, malaise, anorexia, liver big, liver firm, spleen palpable, spiders, ascites, varices, bilirubin, alk phosphate, sgot, albumin, protime, histology. 75 samples have one or more missing value, so we discard them in our analysis. Since this dataset is unbalanced, to avoid over-fitting we have added 30 synthetic entries to the least represented class ("die"), each one obtained selecting randomly one of the original samples, and adding (or subtracting), with probability $p = 0.3$, a small uniform random number to each of its features.

On both datasets, we normalize the input features in the range $[0, 1]$. Also, since both datasets refer to a binary classification problem (positive vs negative diagnosis), for the purpose of classification we set a threshold of 0.5 on the output of the single output node (see the previous section), to discriminate between the two sample classes (corresponding to 0/1 classification values).

4.2   Algorithms

Among the open-source algorithms available in Kimeme, we chose for this comparative study four MOEAs together with a version of Multi-Objective Particle Swarm Optimization used here as control experiment[1]. These algorithms were chosen as they are currently considered the state-of-the-art in multi-objective optimization, and our aim here is to show how general-purpose MOEAs can be used for the automatic design of classifiers. A brief description of the selected algorithms follows, with the related parametrization (for a more thorough explanation of the algorithms and their parameters, please refer to the original papers). All algorithms were configured to use a population of 100 individuals, with stop condition on the number of generations (500).

- Multi-Objective Differential Evolution (MODE). This is a custom multi-objective variant (with elitism) of Differential Evolution (DE) [24], that simply combines with the classic DE mutation/crossover operators the non-dominated sorting and crowding distance mechanisms used in NSGA2 (see below). We set crossover rate $Cr = 0.3$ and scale factor $F = 0.5$.

---

[1] We should note that, technically speaking, MOPSO is not a MOEA, as it is inspired by Swarm Intelligence rather than Evolutionary Algorithms. Nevertheless, for simplicity of notation in the following we will use the wording "MOEAs" to refer generically to all the algorithms tested in this study, including MOPSO.

– Multi-Objective Evolution Strategies (MOES) [25]. This is a multi-objective
  variant of classic Evolution Strategies (ES), an evolutionary algorithm based
  on mutation only. Mutation simply adds to each component of the solution
  a random number drawn from an adaptive distribution. Solutions are then
  ranked, based on their fitness values, to obtain the Pareto front. We set minimum step size $\mu_{min} = 0.01$, initial step size $\mu_{init} = 0.2$, life span $LS = 30$,
  scaling factor $\alpha = 0.2$, and learning rate $\tau = 1$.
– Non-Dominated Sorting Genetic Algorithm-2 (NSGA2) [26]. NSGA2 is
  arguably the most popular algorithm in multi-objective-optimization. It is a
  variant of Genetic Algorithm that uses a non-dominated sorting mechanism
  (to rank solutions based on their dominance level) and a crowding distance
  operator (which preserves a high diversity in the population). We set tournament size $T = 2$, crossover probability $Cr = 0.75$, mutation probability
  $m = 0.05$, selection percentage $s = 0.35$, and exploration factor $e = 0.8$.
– Strength Pareto Evolutionary Algorithm-2 (SPEA2) [27]. SPEA2 relies on
  an archive of boundary solutions and a mechanism for pruning such archive
  along the evolutionary process. Additionally, it incorporates a fine-grained
  fitness assignment strategy based on a nearest-neighbor density estimation
  technique which guides the search more efficiently. We set tournament size
  $T = 2$, crossover probability $Cr = 0.9$, and mutation probability $m = 0.01$.
– Multi-Objective Particle Swarm Optimization (MOPSO) [28]. This is a multi-
  objective variant of Particle Swarm Optimization, that simply combines with
  the classic PSO logics the non-dominated sorting used in NSGA2. The parametrization is the one proposed in [28].

4.3   Minimization of Test Error vs Minimization of Validation Error

The first set of experiments has as main objective the minimization of validation
and test error. Minimizing the validation error allows one to avoid the overfitting
problem, whereas minimizing the test error gives the best performance in terms
of accuracy. We repeat the experiments on each dataset in two conditions, i.e. (1)
one in which the activation function is fixed, a priori and for the entire network,
to one of the following: {Gaussian, Linear, Sigmoid, Sin, Step} and (2) one in
which the activation function is free to vary for each layer, and is chosen by the
MOEA. In the latter case, the activation function is chosen within the following
set of functions: {Bipolar, Competitive, Gaussian, Linear, Log, Ramp, Sigmoid,
Sin, SoftMax, Step, Tanh, Elliott, Symmetric Elliott}; moreover, the MOEA is
allowed to select the same activation function for more than one layer.

In both cases, the neural network topologies are structured as described in
Sect. 3. The number of hidden layers varies in $[1, 3]$, with the first layer having a
variable number of nodes in the range $[1, 255]$, while the size of the second and
the third layers vary in $[0, 255]$, with zero meaning that the layer is not present.
This way, we enforce the condition that the network has at least a hidden layer
made of a single node, while the other two hidden layers might not be present.

We execute the five MOEAs 5 independent times, with different random
seeds, on both datasets in the two conditions. For each algorithm we then

aggregate the Pareto-optimal solutions found at the end of each run, and finally we select the non-dominated solutions among all the optimal solutions found. We report the set of non-dominated solutions obtained by each algorithm on the WBCD in Figs. 2 and 3a, respectively for the case with fixed and variable activation function. As for the HD, due to space limitations we report only the non-dominated solutions obtained with variable activation function, see Fig. 4.

In all figures, the solutions marked with a black square indicate neural networks reaching a full accuracy of 100 %. We should note that, while validation and test error are calculated as Mean Squared Error (MSE) between the expected classification value (0/1) and actual neural network output (ranging in $[0, 1]$, and depending on the output activation function), the accuracy is calculated based on the confusion matrix: (True Positives + True Negatives)/(Total sample size).

The numerical results show that on both datasets and conditions (fixed or variable activation function), the MOEAs obtain several solutions with full accuracy, with no clear superiority of any of the algorithms. Also, the choice of the activation function seems to affect only marginally the performance. Notably, using a variable activation function allowed us to find, on the WBCD, the two full accuracy classifiers with the lowest validation/test error (see the red circle, in Fig. 3a, grey in print). The MOEAs were also successful on the HD, finding numerous full accuracy ANNs, although with a higher validation/test error compared to WBCD (most probably because of the unbalance of the dataset).

To further highlight the potentialities of the MOEA-based method, we report in Tables 1 and 2 a comparison of the best accuracy found in this study against the accuracy obtained in the state-of-art literature, respectively on the WBCD and the HD. We can see that the MOEA-based method tested here is the only one capable of reaching an accuracy value of 100 % on both datasets.

4.4 Minimization of Network Complexity vs Minimization of Validation Error

As an additional experiment, we apply the MOEA-based method to a different formulation of the neural network optimal design, one in which the optimization criteria are the minimization of validation error and network computational complexity. The latter is measured here via the Akaike Information Criterion (AIC) [6], defined as $-2 \cdot \ln(\text{MSE}) + 2k$, where k is the number of weights in the ANNs. This second goal might be important, for instance, in contexts where the classifier ANN must be used in real-time and therefore should be computationally cheap, still guaranteeing robust classification performance.

Due to space limitations, we report only the results on the WBCD (Fig. 3b), but similar considerations apply also to the other dataset. Also, in this case we consider only variable activation functions. Results show that in this case the MOEAs find only one solution with full accuracy, that is associated to the lowest AIC level. On the other hand, ANNs of higher complexity are, unsurprisingly, characterized by a lower validation error but, because of overfitting, none of them is capable of generalizing and obtain full accuracy.

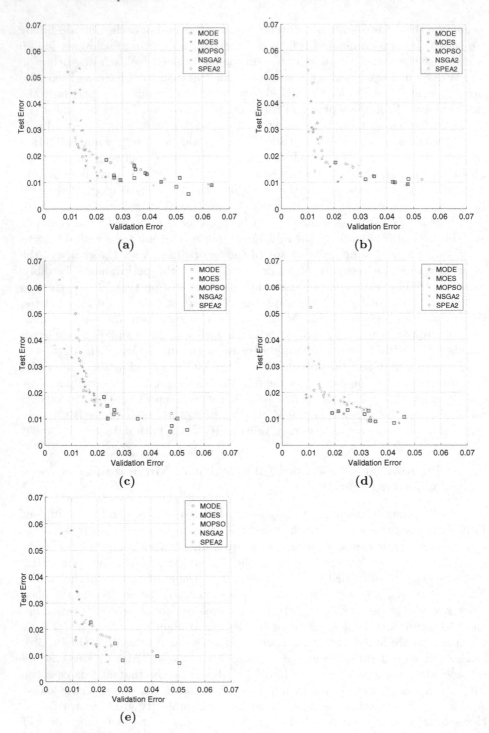

Fig. 2. Non-dominated solutions on the WBCD (with fixed activation function): minimization of validation and test error with Gaussian (a); Linear (b); Sigmoid (c); Sin (d); and Step (e) function.

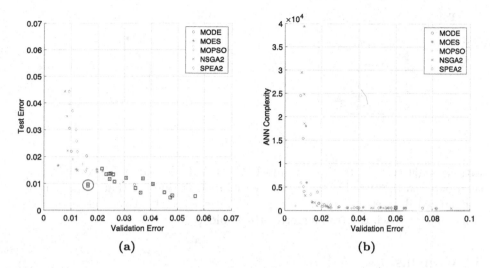

Fig. 3. Non-dominated solutions on the WBCD (with variable activation function): minimization of validation and test error (a); minimization of validation error and Akaike Information Criterion (b).

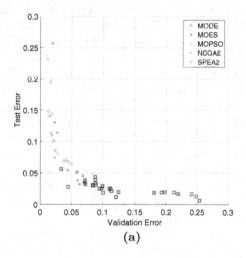

Fig. 4. Non-dominated solutions on the HD (with variable activation function): minimization of validation and test error.

4.5 Execution Times of the Experiments

Finally, we conclude our presentation of the numerical results with a brief analysis of the execution times. On the WBCD, each run of the various MOEAs is executed in approximately 2–6 min. On the HD instead, each run is executed in approximately 2–3 h (except for MOPSO, which takes up to 5 h/run). All tests were executed using 8 threads in parallel on a Linux (Ubuntu 15.04) machine

Table 1. Accuracy on the WBCD

Accuracy	Reference
99.51 %	[29]
99.14 %	[30]
97.8 %	[31]
97.21 %	[32]
100 %	this work

Table 2. Accuracy on the HD

Accuracy	Reference
96.25 %	[33]
94,12 %	[34]
92.9 %	[35]
100 %	this work

with an eight-core i7-5960X CPU and 16 GB RAM. The large difference in run-time between the two datasets can be explained considering that the HD is an unbalanced dataset and the training time on each shuffled version of it takes more time to reach the training error threshold[2].

5   Conclusion

In this paper we have introduced a multi-objective optimization approach for optimally designing and training Artificial Neural Networks used for classification problems. The proposed method leverages several state-of-the-art algorithms provided by Kimeme, an optimization platform available online. We conducted a thorough experimental campaign testing a number of modern multi-objective optimization algorithms, including MOES, MODE, MOPSO, NSGA2 and SPEA2 on two different datasets. Such comparative study was performed on the Breast Cancer Wisconsin Dataset and the Hepatitis dataset from the UCI repository. The aim was to find the non-dominated ANNs minimizing the validation error and the test error, or, alternatively, minimizing at the same time the validation error and the ANN complexity. The latter was measured by means of the Akaike Information Criterion.

All the tested algorithms were able to find, in all conditions on both datasets, several ANNs characterized by 100 % accuracy, improving upon results previously found in the literature. Among the algorithms selected in the study though, we observed a substantial equivalence.

The proposed approach reveals that the automatic design of Artificial Neural Networks by means of multi-objective optimization is a viable solution in the context of complex classification problems. This is especially true when any prior information about the problem at hand is scarce, or not available at all. Furthermore, such an automatic design has a high degree of general purposeness, as it can be easily extended to different classification tasks.

In future studies, we will attempt to test this method on new problems, and we will try to devise novel optimization schemes specifically designed for ML.

---

[2] We should note though, that while in the case of WBCD we used a training error threshold 0.1, in the case of HD we used a threshold of 0.2, to improve the training time and avoid overfitting.

References

1. Baxt, W.G.: Application of artificial neural networks to clinical medicine. Lancet **346**(8983), 1135–1138 (1995)
2. Floyd, C.E., Lo, J.Y., Yun, A.J., Sullivan, D.C., Kornguth, P.J.: Prediction of breast cancer malignancy using an artificial neural network. Cancer **74**(11), 2944–2948 (1994)
3. Bevilacqua, V., Mastronardi, G., Menolascina, F., Pannarale, P., Pedone, A.: A novel multi-objective genetic algorithm approach to artificial neural network topology optimisation: the breast cancer classification problem. In: International Joint Conference on Neural Networks, pp. 1958–1965. IEEE (2006)
4. Fonseca, C., Fleming, P.: Multiobjective genetic algorithms made easy: selection sharing and mating restriction. In: First International Conference on Genetic Algorithms in Engineering Systems: Innovations and Applications, GALESIA 1995, pp. 45–52 (1995)
5. Lichman, M.: UCI Machine Learning Repository (2013). http://archive.ics.uci.edu/ml
6. Akaike, H.: A new look at the statistical model identification. IEEE Trans. Autom. Control **19**(6), 716–723 (1974)
7. Deb, K.: Multi-Objective Optimization Using Evolutionary Algorithms, vol. 16. Wiley, New York (2001)
8. Bevilacqua, V., Costantino, N., Dotoli, M., Falagario, M., Sciancalepore, F.: Strategic design and multi-objective optimisation of distribution networks based on genetic algorithms. Int. J. Comput. Integr. Manufact. **25**(12), 1139–1150 (2012)
9. Ishibuchi, H., Akedo, N., Nojima, Y.: Behavior of multiobjective evolutionary algorithms on many-objective knapsack problems. IEEE Trans. Evol. Comput. **19**(2), 264–283 (2015)
10. Coello, C.A.C.: Multi-objective evolutionary algorithms in real-world applications: some recent results and current challenges. In: Greiner, D., Galván, B., Périaux, J., Gauger, N., Giannakoglou, K., Winter, G. (eds.) Advances in Evolutionary and Deterministic Methods for Design, Optimization and Control in Engineering and Sciences, pp. 3–18. Springer, New York (2015)
11. Bevilacqua, V., Mastronardi, G., Piscopo, G.: Evolutionary approach to inverse planning in coplanar radiotherapy. Image Vision Comput. **25**(2), 196–203 (2007)
12. Menolascina, F., Bellomo, D., Maiwald, T., Bevilacqua, V., Ciminelli, C., Paradiso, A., Tommasi, S.: Developing optimal input design strategies in cancer systems biology with applications to microfluidic device engineering. BMC Bioinform. **10**(S-12), 1–4 (2009)
13. Whitley, D., Starkweather, T., Bogart, C.: Genetic algorithms and neural networks: optimizing connections and connectivity. Parallel Comput. **14**(3), 347–361 (1990)
14. Ilonen, J., Kamarainen, J.K., Lampinen, J.: Differential evolution training algorithm for feed-forward neural networks. Neural Process. Lett. **17**(1), 93–105 (2003)
15. Leung, F.H., Lam, H.K., Ling, S.H., Tam, P.K.: Tuning of the structure and parameters of a neural network using an improved genetic algorithm. IEEE Trans. Neural Netw. **14**(1), 79–88 (2003)
16. Bhardwaj, A., Tiwari, A.: Breast cancer diagnosis using genetically optimized neural network model. Expert Syst. Appl. **42**(10), 4611–4620 (2015)

17. Bevilacqua, V., Brunetti, A., de Biase, D., Tattoli, G., Santoro, R., Trotta, G.F., Cassano, F., Pantaleo, M., Mastronardi, G., Ivona, F., et al.: A P300 clustering of mild cognitive impairment patients stimulated in an immersive virtual reality scenario. In: Intelligent Computing Theories and Methodologies, pp. 226–236. Springer (2015)

18. Bevilacqua, V., Salatino, A.A., Di Leo, C., Tattoli, G., Buongiorno, D., Signorile, D., Babiloni, C., Del Percio, C., Triggiani, A.I., Gesualdo, L.: Advanced classification of alzheimer's disease and healthy subjects based on EEG markers. In: International Joint Conference On Neural Networks, pp. 1–5. IEEE (2015)

19. Bevilacqua, V., Tattoli, G., Buongiorno, D., Loconsole, C., Leonardis, D., Barsotti, M., Frisoli, A., Bergamasco, M.: A novel BCI-SSVEP based approach for control of walking in virtual environment using a convolutional neural network. In: International Joint Conference On Neural Networks, pp. 4121–4128. IEEE (2014)

20. Riedmiller, M., Braun, H.: RPROP-a Fast Adaptive Learning Algorithm. In: Proceedings of ISCIS VII, Universitat (1992)

21. Cyber Dyne Srl: Kimeme. http://cyberdynesoft.it/

22. Iacca, G., Mininno, E.: Introducing kimeme, a novel platform for multi-disciplinary multi-objective optimization. In: Rossi, F., et al. (eds.) WIVACE 2015. CCIS, vol. 587, pp. 40–52. Springer, Heildelberg (2016). doi:10.1007/978-3-319-32695-5_4

23. Heaton, J.: Programming Neural Networks with Encog 2 in Java (2010)

24. Storn, R., Price, K.: Differential evolution-a simple and efficient heuristic for global optimization over continuous spaces. J. Global Optim. **11**(4), 341–359 (1997)

25. Beyer, H.G., Arnold, D.V.: Theory of evolution strategies - a tutorial. In: Kallel, L., Naudts, B., Rogers, A. (eds.) Theoretical Aspects of Evolutionary Computing, pp. 109–133. Springer, New York (2001)

26. Deb, K., Pratap, A., Agarwal, S., Meyarivan, T.: A fast and elitist multiobjective genetic algorithm: NSGA-II. IEEE Trans. Evol. Comput. **6**(2), 182–197 (2002)

27. Zitzler, E., Laumanns, M., Thiele, L., Zitzler, E., Zitzler, E., Thiele, L., Thiele, L.: SPEA2: Improving the Strength Pareto Evolutionary Algorithm (2001)

28. Wickramasinghe, U., Li, X.: Choosing leaders for multi-objective PSO algorithms using differential evolution. In: Li, X., et al. (eds.) SEAL 2008. LNCS, vol. 5361, pp. 249–258. Springer, New York (2008)

29. Akay, M.F.: Support vector machines combined with feature selection for breast cancer diagnosis. Expert Syst. Appl. **36**(2), 3240–3247 (2009)

30. Şahan, S., Polat, K., Kodaz, H., Güneş, S.: A new hybrid method based on fuzzy-artificial immune system and k-nn algorithm for breast cancer diagnosis. Comput. Biol. Med. **37**(3), 415–423 (2007)

31. Pena-Reyes, C.A., Sipper, M.: A fuzzy-genetic approach to breast cancer diagnosis. Artif. Intell. Med. **17**(2), 131–155 (1999)

32. Setiono, R., Liu, H.: Symbolic representation of neural networks. Computer **29**(3), 71–77 (1996)

33. Sartakhti, J.S., Zangooei, M.H., Mozafari, K.: Hepatitis disease diagnosis using a novel hybrid method based on support vector machine and simulated annealing (SVM-SA). Comput. Methods Programs Biomed. **108**(2), 570–579 (2012)

34. Polat, K., Güneş, S.: Prediction of hepatitis disease based on principal component analysis and artificial immune recognition system. Appl. Math. Comput. **189**(2), 1282–1291 (2007)

35. Bascil, M.S., Oztekin, H.: A study on hepatitis disease diagnosis using probabilistic neural network. J. Med. Syst. **36**(3), 1603–1606 (2012)

Novel Algorithm for Efficient Distribution of Molecular Docking Calculations

Luigi Di Biasi[1,2], Roberto Fino[1], Rosaura Parisi[1], Lucia Sessa[1],
Giuseppe Cattaneo[2], Alfredo De Santis[2], Pio Iannelli[1],
and Stefano Piotto[1(✉)]

[1] Department of Pharmacy, University of Salerno, Via Giovanni Paolo II,
132-84084 Fisciano, SA, Italy
piotto@unisa.it
[2] Department of Informatics, University of Salerno, Via Giovanni Paolo II,
132-84084 Fisciano, SA, Italy

Abstract. Molecular docking is a computational method to study the formation of intermolecular complexes between two molecules. In drug discovery, it is employed to estimate the binding between a small ligand (the drug candidate), and a protein of known three-dimensional structure. Docking is becoming a standard part of workflow in drug discovery. Recently, we have used the software VINA, a de facto standard in molecular docking, to perform extensive docking analysis. Unfortunately, performing a successful blind docking procedure requires large computational resources that can be obtained by the use of clusters or dedicated grid. Here we present a new tool to distribute efficiently a molecular docking calculation onto a grid changing the distribution paradigm: we define portions on the protein surface, named hotspots, and the grid will perform a local docking for each region. Performance studies have been conducted via the software GRIMD.

1 Introduction

Drug discovery is a time-consuming, risky, and expensive process. To shorten the research cycle and to lower the failure rate, Computer-Aided Drug Design is applied in the early drug discovery phases. Molecular docking is one of the most popular strategies to evaluate the drugability of a molecule. An efficient search of the best binding interaction of a ligand is extremely computationally demanding, and existing software suffer several limitations. As result, the predictive ability of docking software is severely limited, especially for blind docking. Different approaches have been used by different research groups to identify the receptor binding site and to estimate all the contributions to the total binding energy of the ligand – receptor complex. No commercially available or free-to-use software for molecular docking consider the importance of conserved sequence in proteins. Often the active site of the receptor is unknown, so the only option left to identify all the possible binding sites on the receptor is to extend the search box to all the receptor, a type of molecular docking also known as blind docking, reducing the accuracy of the procedure and leading to an increase of computing times. The existing software are, from a computational point of view, extremely inefficient. In fact, programs like Vina [1] can generate thousands of

F. Rossi et al. (Eds.): WIVACE 2015, CCIS 587, pp. 65–74, 2016.
DOI: 10.1007/978-3-319-32695-5_6

poses with a small coverage of the conformational space. What is worst is that the best pose is often rejected even after an exhaustive pose generation. This means that the program found a pose close to the experimental one but the scoring function poorly evaluated it. The pose ranking is based on a calculated binding energy that shows a poor correlation with experimental binding energies. For this reason, one cannot simply rely on massive calculation of an astronomic number of poses, because the ranking functions are not properly calibrated. During extensive docking analysis, we observed that conserved residues often lie on binding sites [2, 3]. Our idea was to drive ligands toward conserved regions on the surface adding an extra term to the force field. We decided to use the software Vina because it is efficient and open source. We could observe that, in most cases, binding sites lie on conserved portion on the protein surface [2, 3]. The opposite in not always true, so we can assume that the presence of conserved residues is a necessary but not sufficient condition to predict a binding site. Conserved residues are rarely isolated. Normally, a binding site can be made of several spatially closed but non-adjacent residues.

2   Experimental Part

We define as hotspot (HS) the barycenter of spatially related conserved residues. The conserved regions can be easily obtained by multiple sequence analysis, but an easier way consists in downloading essential information from the server PDBFinder [4]. The distance of a pose from the HS is then used to modify the Vina function.

The poses that satisfied the Vina criteria are checked out in terms of distance from the HS. The binding energy is than modified according to Eq. (1), adding a term that depends on the minimal distance (d) between the ligand barycenter and the nearest HS. The new energy takes into account also the conservation value of the residue (conservation weight, C_W):

$$E_{QN} = E_{QN}^0 + \frac{E_{QN}^0 C_w}{d^2} \tag{1}$$

These values of E_{QN} (quasi-Newton energy) are saved in the Prop channel and used to train the genetic algorithm.

```
void quasi_newton::operator()(model& m, const
precalculate& p, const igrid& ig, output_type& out,
change& g, const vec& v) const { // g must have correct
size
        quasi_newton_aux aux(&m, &p, &ig, v);
        fl res = bfgs(aux, out.c, g, max_steps, aver-
age_required_improvement, 10);
        fl hsval = m.eval_conservation(out.coords,out);
        out.yadaRankProp = res*hsval;
        out.dist = hsval;
        out.e = res;
        }
```

In red is highlighted where the calculation is called in the original quasi_newton. cpp Vina file. The definition of the function eval_conservation is in the file model.cpp.

```
fl model::eval_conservation(vecv poseCoords,output_type&
out)
{
        //find nearest hot spot
        if (poseCoords.size()==0)
                return 0;
        // for each hot spot check which of them is close
to current ligand
        int j;
        int hsmin = 0;
        fl min_distance =
std::numeric_limits<float>::max();;
        for (j=0;j<yada_hscoords.size();j++)
        {
                fl r;
                int k;
                r = 0;
                int c;
                for (k=0;k<=poseCoords.size();k++)
                        for (c=0;c<=3;c++)
                                r+= pow(poseCoords[k][c]
-yada_hscoords[j][c],2);
                r = sqrt(r);
                if (r<min_distance)
                {
                        min_distance = r;
                        hsmin=j;
                }
        }
        // nearest hotspot found. Looking for reference
atom.
        int k;
        min_distance = std::numeric_limits<float>::max();
        for (k=0;k<=poseCoords.size();k++)
        {
                // check distance between atom and
'reference atom' in nearest
                // hotspot
                int c;
                float r;
                r=0;
                for (c=0;c<=3;c++)
                        r+= pow(poseCoords[k][c] -
yada_hscoords[hsmin][c],2);
                r = sqrt(r);
                if (r<min_distance)
                {
                        min_distance = r;
                }
        }
        // save nearest atom
        vec
hsrefatompos(yada_hscoords[hsmin][0],yada_hscoords[hsmin]
[1], yada_hscoords[hsmin][2]);
        out.hsrefatomcoords = hsrefatompos;
        // return prop
        return
pow(yada_hscoords[hsmin].hsval,2)*(1/min_distance);
}
```

When we want to explicitly consider the presence of water molecules, the program checks if there is space to introduce an oxygen atom at 2.9 Å from electronegative atoms in the ligand. If so, the QN function is called with and without oxygen and only the pose with the maximum QN is chosen.

The choice of poses is made with a traditional Metropolis approach. Metropolis algorithm is a Markov chain Monte Carlo method for obtaining a sequence of random poses from a probability distribution for which direct sampling is difficult. When the energy results are to be higher, the new conformation will be accepted or rejected if an acceptance probability law

$$P = e^{\left[-\frac{E_2 - E_1}{k_b T}\right]}$$

(2)

is randomly satisfied, where T is temperature and kB the Boltzmann's constant. The acceptance condition is verified if generating a pseudo-random number u, uniformly distributed between 0–1, will result $u < P$.

The following changes have been introduced in metropolis.cpp.

```
bool metropolis_accept(fl old_f, fl new_f, fl tempera-
ture, rng& generator,fl dist,output_type& m ) {
        if (dist>5) return false;
        if(new_f < old_f) return true;
        const fl acceptance_probability = std::exp((old_f
- new_f) / temperature);
        // flip coin here,
        return random_fl(0, 1, generator) < ac-
ceptance_probability;
}
```

The initial idea was to increase the acceptance probability nearby conserved regions in order to drive the ligands toward those sites. The acceptance rate was therefore modified as:

$$AccProb = AccProb \cdot \frac{1}{\left(distance_{pose-hotspot}\right)^2}$$

(3)

The distance is obtained from the routine Quasi_Newton to decrease the computing time. The instruction m.metEnable, reads the flag to tell the program when use the modified Metropolis.

```
bool metropolis_accept(fl old_f, fl new_f, fl tempera-
ture, rng& generator,fl dist,output_type& m ) {
        if (dist>5) return false;
        if(new_f < old_f) return true;
        fl yada_increase = 0;
        if (m.metenable)
            yada_increase = 1/pow(m.hsdistance,2)

        const fl acceptance_probability = std::exp((old_f
- new_f) / temperature) + yada_increase;
return random_fl(0, 1, generator) < ac-
ceptance_probability;
}
```

The flowchart representing the implementation of HS in the original Vina code is shown in Fig. 1.

2.1 Improving the Scoring Function

One of the typical problems with docking software, and Vina makes no exception, is that the pose ranking is made in terms of energy. Vina uses a semi-empiric calculation of the pose energy. Unfortunately, the calculation of free energy is far from being optimal and, consequently, the ranking process is poor. This means that the best pose, i.e. the one with the minimum RMSD from experimental data, is not the first in rank. We have performed an extensive genetic algorithm study to improve the ranking. As result, in 95 % of the cases, the best pose is the one with the highest score.

2.2 Porting on Grimd

Grimd is a software that can easily create a computer grid [5]. Grimd can chunk a complex calculation in a number of smaller jobs. The jobs are sent to available PCs (slaves) and, after completion, the most relevant results are collected and made available via web interface. The Master, which is a dedicated machine, runs a program that takes care of the input data partitioning, the scheduling, the tasks execution across a set of machines (dynamically updated), the handling of machine failures, and the managing of the required inter-machine communication. The Master implements a basic authentication mechanism when a new slave subscribes to the "Grimd Network", managing communication privacy through channel encryption (a sort of VPN) and client-side strong authentication through session key negotiation. The distributed grid was already successfully applied for a wide range of applications [6–10]. Grimd was used to perform a flexible ligand-flexible receptor docking encoding the conformational spaces of molecules through a protocol of molecular dynamics, followed by the generation of an ensemble of rotamers. These conformational subspaces can be built to span a range of conformations important for the biological activity of a protein.

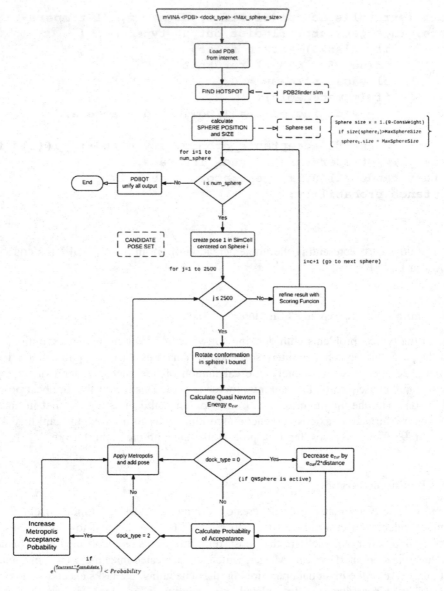

Fig. 1. Flowchart of the implementation of HS on Vina code

A variety of motions can be combined, ranging from domains moving as rigid bodies or backbone atoms undergoing normal mode-based deformations, to side chains assuming rotameric conformations. In addition, Grimd can be easily used for the screening of several receptors against a large library of ligands. Because of the underlying architecture, Grimd is not limited to docking or molecular dynamics, but it has been also applied to extend coarse grain dynamics [11, 12], to distribute quantum mechanical calculations [13, 14], and to improve of orders of magnitude the speed of

Monte Carlo simulations [15]. The concept of hotspot permitted a straightforward distribution on a grid. In fact, each hotspot can be computed on a different node of the grid. To run a full blind docking calculation, it is necessary to send three files to the Master: the receptor and the ligand in pdbqt format, and a text file containing the conservation string. An example of the text file is:

```
##EXPLOSION(1) = WRITE_RANGE[30,60,1] @ENDEXP
1au2_ligand.pdbqt 1au2_receptor.pdbqt
89357988459386588598399899799448689753134684754994969899
138399399473389497227297658439891625419274322257232661475
254925832886136877767874125955933989263273333669687589992
326245977899983299349962757634717888832878926
#min_range = '@EXPLOSION(1)@'
#hotspot_pos = @HS@
#exit
```

Once the Master received the job request, each HS is assigned to a different Slave. Once the local calculation is completed, the Slaves create two files: a file with results and a log file. These files are sent back to the Master that reduces the information into a global result file, sorts the poses using the binding energy and print out one or more poses. No changes are necessary on the Master to distribute the calculation. The flowchart representing the porting of Vina on Grimd is shown in Fig. 2.

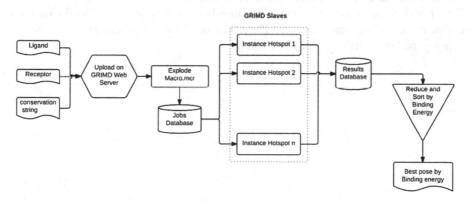

Fig. 2. Flowchart of VINA porting on GRIMD

3   Results

3.1   Docking Validation

We tested the accuracy of the new approach on a customized version of the Aspex list of PDBs available in literature. The validation of a docking software is always a critical task. Several works already discussed this point [3]. Among others, two aspects appear to be critical: the choice of the validation data set, and the position (and dimension) of the docking box. In fact, it is evident that an ad hoc choice of the proteins to be docked can give the illusion of fantastic performance. Few publications offer a systematic comparison of software performances. The validation did not use any prior information on the docking box, or ligand orientation, nor had we used a particular (and benevolent) pdb validation set. The evaluation of was performed on a set of 180 proteins and the results compared to Vina, a de facto standard in molecular docking. The docking procedure was total blind docking, 250 runs, Amber03 ff, no water molecules. We have considered two aspects in blind docking: the goodness of the first pose in terms of RMSD between the docked pose and the experimental data, the free energy of binding and the execution time.

3.2   Grid Tests

Standard procedures were followed to set up the virtual docking. AutoDock requires that the ligands and receptor be formatted in pdbqt files. This format is similar to a PDB file and also has charge and AutoDock atom-type information. These files can be created with the AutoDockTools (ADT) [16] interface or with scripts provided with the software. ADT is a graphical interface provided with the AutoDock software and can be used to carry out serial docking jobs, prepare files, and analyze results. The provided scripts were used to add partial charges to each atom, merge nonpolar hydrogens with the heavy atoms to which they are covalently bound and determine the AutoDock atom types of each atom for all the ligands and decoys. Using HSs, a docking calculation with Vina can be easily distributed on several machines.

In Fig. 3, it is shown the computing time for the system 1 fkg distributed on a number of nodes between 1 and 10. The behavior of the net is almost linear demonstrating a perfect distribution of the docking.

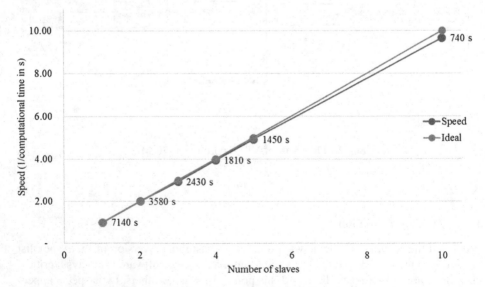

Fig. 3. Running time as function of slave number

4   Conclusions

We presented here a modification of Vina that permitted to increase the accuracy of docking as well as a load distribution on a dedicated grid that permitted a drastic reduction of the computational time. The pose generation algorithms and the scoring function for pose ranking have been modified in order to consider the conservation of residues in the protein sequence. To assign the conservation weights to each residue we used a customized version of the HSSP database [4]. In correspondence to each conserved region, the program places new local search boxes. The size of these boxes is inversely depending on the conservation of the residue: the highest the conservation, the smallest is the size. The choice of the best pose follows a completely novel approach. We have used Genetic Algorithms (GA) to develop a scoring function that takes into account force field related energy as well as the distance of the ligand from conservation regions. According to this, our new ranking function allows the final user to pick the best pose after the molecular docking with a better accuracy and reliability. Finally, the introduction of hotspots, i.e. highly conserved residues in a protein, permitted a straightforward and efficient distribution onto a dedicated grid.

Acknowledgments. This work was partially supported by the "Data-Driven Genomic Computing (GenData 2020)" PRIN project (2013–2015), funded by the Italian Ministry of the University and Research (MIUR).

References

1. Trott, O., Olson, A.J.: AutoDock Vina: improving the speed and accuracy of docking with a new scoring function, efficient optimization, and multithreading. J. Comput. Chem. **31**(2), 455–461 (2010)
2. de Vries, S.J., van Dijk, A.D., Bonvin, A.M.: WHISCY: What information does surface conservation yield? Application to data-driven docking. Proteins Struct. Funct. Bioinf. **63**(3), 479–489 (2006)
3. Ouzounis, C., Pérez-Irratxeta, C., Sander, C., Valencia, A.: Are binding residues conserved? In: Pacific Symposium on Biocomputing, pp. 401–412 (1997)
4. Hooft, R.W., Sander, C., Scharf, M., Vriend, G.: The PDBFINDER database: a summary of PDB, DSSP and HSSP information with added value. Comput. Appl. Biosci. CABIOS **12**(6), 525–529 (1996)
5. Piotto, S., Di Biasi, L., Concilio, S., Castiglione, A., Cattaneo, G.: GRIMD: distributed computing for chemists and biologists. Bioinformation **10**(1), 43 (2014)
6. Concilio, S., Bugatti, V., Neitzert, H.C., Landi, G., De Sio, A., Parisi, J., Piotto, S., Iannelli, P.: Zn-complex based on oxadiazole/carbazole structure: synthesis, optical and electric properties. Thin Solid Films **556**, 419–424 (2014)
7. Lopez, D.H., Fiol-deRoque, M.A., Noguera-Salvà, M.A., Terés, S., Campana, F., Piotto, S., Castro, J.A., Mohaibes, R.J., Escribá, P.V., Busquets, X.: 2-Hydroxy arachidonic acid: a new non-steroidal anti-inflammatory drug. PLoS ONE **8**(8), e72052 (2013)
8. Piotto, S., Concilio, S., Bianchino, E., Iannelli, P., López, D.J., Terés, S., Ibarguren, M., Barceló-Coblijn, G., Martin, M.L., Guardiola-Serrano, F.: Differential effect of 2-hydroxyoleic acid enantiomers on protein (sphingomyelin synthase) and lipid (membrane) targets. Biochim. Biophys. Acta (BBA)-Biomembr. **1838**(6), 1628–1637 (2014)

9. Piotto, S., Trapani, A., Bianchino, E., Ibarguren, M., López, D.J., Busquets, X., Concilio, S.: The effect of hydroxylated fatty acid-containing phospholipids in the remodeling of lipid membranes. Biochim. Biophys. Acta (BBA)-Biomembr. **1838**(6), 1509–1517 (2014)

10. Scrima, M., Di Marino, S., Grimaldi, M., Campana, F., Vitiello, G., Piotto, S.P., D'Errico, G., D'Ursi, A.M.: Structural features of the C8 antiviral peptide in a membrane-mimicking environment. Biochim. Biophys. Acta (BBA)-Biomembr. **1838**(3), 1010–1018 (2014)

11. Caracciolo, G., Piotto, S., Bombelli, C., Caminiti, R., Mancini, G.: Segregation and phase transition in mixed lipid films. Langmuir **21**(20), 9137–9142 (2005)

12. Piotto, S., Concilio, S., Mavelli, F., Iannelli, P.: Computer simulations of natural and synthetic polymers in confined systems. Macromol. Symp. **286**(1), 25–33 (2009)

13. Piotto, S., Nesper, R.: CURVIS: a program to study and analyse crystallographic structures and phase transitions. J. Appl. Crystallogr. **38**(1), 223–227 (2005)

14. Acierno, D., Amendola, E., Bugatti, V., Concilio, S., Giorgini, L., Iannelli, P., Piotto, S.: Synthesis and characterization of segmented liquid crystalline polymers with the azo group in the main chain. Macromolecules **37**(17), 6418–6423 (2004)

15. Piotto, S., Mavelli, F.: Monte Carlo simulations of vesicles and fluid membranes transformations. Orig. Life Evol. Biosph. **34**(1–2), 225–235 (2004)

16. Morris, G.M., Huey, R., Lindstrom, W., Sanner, M.F., Belew, R.K., Goodsell, D.S., Olson, A.J.: AutoDock4 and AutoDockTools4: automated docking with selective receptor flexibility. J. Comput. Chem. **30**(16), 2785–2791 (2009)

Modelling and Simulation of Artificial and Biological Systems

Simulating Bacteria-Materials Interactions via Agent-Based Modeling

Maria A. Bonifacio[1], Stefania Cometa[2], and Elvira De Giglio[1(✉)]

[1] Chemistry Department, University "Aldo Moro", Via Orabona 4, 70125 Bari, Italy
elvira.degiglio@uniba.it
[2] Jaber Innovation s.r.l, Via Calcutta 8, 00100 Rome, Italy

Abstract. This work reports the outcomes of *in silico* simulations of the interactions between *S. aureus* bacteria and an antibacterial polymeric coating developed onto titanium substrates. The aim of the theoretical analysis is to develop a computational approach suitable of predicting the effective amount of antibacterial agents to load onto the polymeric coating in order to prevent titanium implants infections and at the same time to minimize cytotoxicity. The simulations results will be contrasted with experimental data.

Keywords: Antibacterial activity · Titanium implants · Infections · In silico modeling

1 Introduction

In biomaterials science, the development of innovative antibacterial surfaces is one of the most active research fields: infections, indeed, often cause orthopedic and dental implant failure. Implant associated infections and antibiotic resistance are still rising, with *S. aureus* and coagulase-negative *staphylococci* accounting for 45–55 % of infections [1–4]. Therefore, antibacterial biomaterials are essential to improve *in vivo* implant duration [5–8]. Antibacterial agents, loaded and released at the implant site, are an effective approach to address this urgent issue. Silver ions are among the most promising non-conventional antimicrobial agents, although their mechanism of action, as well as their toxicity, are controversial and poorly understood [3].

Wet laboratory experiments on cell-biomaterial interactions are often combined with computational studies because they are cost effective and time saving. These interesting advantages suggest the opportunity for *in silico* methods to become, in the next future, valuable tools to design, test and optimize functional biomaterials [9–11].

In this research field, *agent-based models* are a class of computational techniques that appear to be highly promising since they have been largely applied to describe the macroscopic properties of dynamical systems characterized by heterogeneity and self-organization [12]. They simulate the simultaneous operations and interactions of *agents*, the essential elements describing the studied systems, in an attempt to re-create and predict the appearance of complex phenomena. In fact, *in silico* complex behaviours

© Springer International Publishing Switzerland 2016
F. Rossi et al. (Eds.): WIVACE 2015, CCIS 587, pp. 77–82, 2016.
DOI: 10.1007/978-3-319-32695-5_7

to a higher (macro) level can be observed starting from setting simple rules of agent-agent and agent-environment interactions. This bottom-up method has been exploited to investigate the dynamics of complex populations at different levels, from the sociological to the cellular and atomic scale [13]. On the other hand, biological systems are particularly challenging for classical computational methods, because the information flux between single elements over time, as well as between the elements and the surrounding environment, must be taken into account. When also the spatial distribution plays a crucial role, such as for biomaterials, the agent-based approach is particularly suitable: agents are free to move and act independently from each other [14]. Moreover, in the specific case of antibacterial surfaces, agent-based models enable the study of the effects of surface stimuli on bacterial populations [15].

In this work, an agent-based modeling approach has been adopted to simulate *S. aureus* interactions with a titanium substrate loaded with silver ions, in order to choose the least effective amount of the antibacterial agent. The outcomes of the *in silico* simulations are compared with preliminary experimental observations.

2   Materials and Methods

2.1   Preparation of the Antibacterial Surface

The polymeric coating has been synthesized on 1 cm^2 titanium substrates according to an electrochemical procedure previously described and different amounts of silver ions per unit surface area have been loaded onto the polymeric coating as antibacterial agents [16]. All electrochemical experiments have been carried out using a PAR Versa STAT 4 potentiostat - galvanostat (Princeton Applied Research). The coated titanium samples have been sterilized by UV treatment before *in vitro* antibacterial evaluation.

2.2   *In vitro* Antibacterial Activity

All the experiments have been performed in triplicate according to ISO 22196:2011. *S. aureus* has been grown in Luria Bertani Broth (LB) (Difco, Detroit, MI, USA).

A suspension of 10^4 bacteria has been incubated onto each UV sterilized titanium substrate (and onto sterile substrates as control) for 24 h at 37 °C. Then, bacteria have been serially diluted and plated on agar. After 24 h at 37 °C bacterial viability has been calculated setting to 100 % the control growth (bacteria grown onto sterile substrates) and comparing bacterial number onto the bare coatings and onto silver-loaded coatings.

2.3   *In silico* Simulations

The NetLogo platform has been used to simulate the time behaviour of *S. aureus* bacterial population deposited on a titanium surface coated by a polymeric layer loaded by different amounts of silver ions [17].

Simulation consists in a random walk of bacteria on a squared surface discretized in 33 × 33 lattice. Each node of the lattice corresponds to a bacterium binding site. Since the cell diameter of *S. aureus* is approximately 0.6 μm, by assuming a spherical shape, the surface area of a bacterium is around 0.3 μm^2 and therefore 33 × 33 lattice corresponds to a surface of 300 μm^2 with about 10^3 binding sites. The simulation time is expressed as clock ticks, that are time steps in which all bacteria can move, die or reproduce. The length of the clock tick has been estimated equal to 10 min and the simulations last for 144 ticks, corresponding to the *in vitro* test duration (24 h).

At each step of the simulation bacteria are allowed to do:

– a random movement, that is to move in an adjacent binding site by chance. A positive score of 2, called *energy*, has been assigned to bacteria (turtles) moved on a polymer patch. The latter changes colour after bacterial passage;
– bacteria standing onto randomly-positioned silver ion patches get killed and disappear from the lattice;
– when bacteria have reached the division energy, equal to 250, they reproduce (*hatch*) creating two new-born bacteria, each having a starting energy of 25.

The output of the simulation consists in a plot of the total number of bacteria against time, combined with a 3D animation showing the time evolution of the bacterial population on the coated titanium surface. In Fig. 1, three snapshots of this animation have been reported, that enhance the visualization of the system's behaviour.

In order to easily simulate samples containing different amounts of silver ions, a slider bar has been inserted in the graphical user interface of the NetLogo model, allowing the rapid tuning of the system without any code changes. With the same purpose, other sliders (starting bacterial number and energy gained from polymer) have been added to create a more user-friendly interface.

The antibacterial activity of the simulated substrate is assumed to reach the lower threshold of effectiveness when the initial bacterial population is not allowed to grow and fluctuates within 10 units around the initial value.

3 Results and Discussion

The antibacterial activity threshold for silver-embedded polymer coating is reported in Table 1. It has been estimated by *in vitro* experiments and refers to a surface coverage of silver ions equal to 0.9 %. Loading samples with lower amounts of silver results in limited activity, while a 3.6 % surface coverage of silver ions proves a higher efficiency *in vitro*. Conversely, bacteria grow with no significant difference onto the bare polymeric matrix compared to the control culture (viability 103 ± 1 %). As further control, bacterial growth onto uncoated titanium has been studied and no differences have been seen compared to the control culture (data not shown).

The *in silico* simulation outcomes are in agreement with the *in vitro* observations. When silver ions are absent on the polymer-coated titanium surface after a 144 tick simulation, i.e. 24 h *in vitro*, the entire lattice is covered by bacteria and this corresponds to the viability observed by *in vitro* experiments. On the other hand when silver ions are present with a cover percentage of 0.9 %, i.e. setting 10 lattice sites as silver patches, starting from a bacterial population of 2 % coverage (20 bacteria on the 33 × 33 binding site lattice), the bacterial population do not grow indefinitely but it fluctuates around its starting value, see Fig. 1. According to the *in vitro* experiments, a higher amount of silver ions exhibits a stronger antimicrobial activity and the bacterial population is highly reduced also *in silico*. Conversely, the system with the lower silver amount than 0.9 % coverage fails to maintain bacterial population into the starting range. Indeed, it has been possible to simulate the performances of the titanium coating loaded with different amounts of silver ions, reproducing *in vitro* experiments.

Table 1. In *vitro* antibacterial activity of the proposed coating

Surface coverage of silver ions	Bacterial viability
0.2 ± 0.1 %	18 ± 1 %
0.9 ± 0.1 %	2 ± 1 %
3.6 ± 0.1 %	0
Polymer only	103 ± 1 %

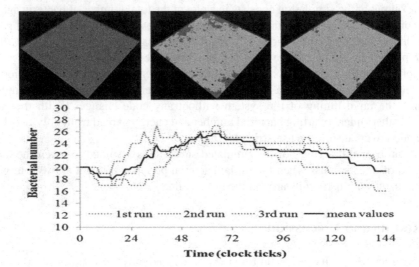

Fig. 1. The lower plot reports the bacterial number against time obtained as outcomes of three different simulations (dotted lines) all of them starting from an initial population of 20 bacteria on a polymer-coated titanium surface loaded with silver ions at 0.9 % coverage; the black line is the average time course of the bacterial population. On top, three different snapshots are shown taken at different simulation times (clock tick 0, 72 and 144), during the run reported in green in the upper plot; initial bacteria are reported as grey dots while new-born bacteria are in green, dark squares represent silver ions.

4   Conclusions

The proposed simulation describes agent-environment interactions in agreement with *in vitro* microbiological tests. As a further improvement, it would be interesting to study the same bacteria-material interactions for a time longer than 24 h, looking for an agreement between *in vitro* and *in silico* observations. The simulations would be closer to reality taking into account *S. aureus* ability to aggregate forming *grape* clusters, or to produce biofilms to attach to substrates and shield from antimicrobials.

Even though very simple, this computational approach could help reducing time and costs of wet lab experiments, improving the antibacterial performances of current orthopedic and dental implants.

References

1. Drancourt, M., Stein, A., Argenson, J.N., Roiron, R., Groulier, P., Raoult, D.: Oral treatment of Staphylococcus spp. infected orthopaedic implants with fusidic acid or ofloxacin in combination with rifampicin. J. Antimicrob. Chemother. **39**(2), 235–240 (1997)
2. Bisno, A.L., Waldvogel, F.A.: Infected orthopedic prostheses. In: Infections Associated With Indwelling Medical Devices, pp. 111–127 (1989)
3. Jung, W.K., Koo, H.C., Kim, K.W., Shin, S., Kim, S.H., Park, Y.H.: Antibacterial activity and mechanism of action of the silver ion in Staphilococcus aureus and Escherichia coli. Appl. Environ. Microbiol. **74**(7), 2171–2178 (2008)
4. Montanaro, L., Speziale, P., Campoccia, D., Ravaioli, S., Cangini, I., Pietrocola, G., Giannini, S., Arciola, C.R.: Scenery of Staphylococcus implant infections in orthopedics. Future Microbiol. **6**(11), 1329–1349 (2011)
5. Gallo, J., Holinka, M., Moucha, C.S.: Antibacterial surface treatment for orthopaedic implants. Int. J. Mol. Sci. **15**(8), 13849–13880 (2014)
6. Busscher, H.J., van der Mei, H.C., Subbiahdoss, G., Jutte, P.C., van den Dungen, J.J., Zaat, S.A., Schultz, M.J., Grainger, D.W.: Biomaterial-associated infection: locating the finish line in the race for the surface. Sci. Transl. Med. **26**(4), 153–163 (2012)
7. Mattioli-Belmonte, M., Cometa, S., Ferretti, C., Iatta, R., Trapani, A., Ceci, E., Falconi, M., De Giglio, E.: Characterization and cytocompatibility of an antibotic/chitosan/cyclodextrins nanocoating on titanium implants. Carb. Pol. **110**, 173–182 (2014)
8. Moriarty, T.F., Zaat, S.A.J., Busscher, H.J.: Biomaterials Associated Infection. Immunological Aspects and Antimicrobial Strategies. Springer, Heildelberg (2013). ISBN: 978-1-4614-1030-0
9. Shamloo, A., Mohammadaliha, N., Mohseni, M.: Integrative utilization of microenvironments, biomaterials and computational techniques for advanced tissue engineering. J. Biotechnol. **212**, 71–89 (2015)
10. Sanz-Herrera, J.A., Reina-Romo, E.: Cell-biomaterial mechanical interaction in the framework of tissue engineering: insights, computational modeling and perspectives. Int. J. Mol. Sci. **12**(11), 8217–8244 (2011)
11. Gronau, G., Krishnaji, S.T., Kinahan, M.E., Giesa, T., Wong, J.Y., Kaplan, D.L., Buehler, M.J.: A review of combined experimental and computational procedures for assessing biopolymer structure-process-property relationships. Biomaterials **33**(33), 8240–8255 (2012)
12. Axelrod, R.: The Complexity of Cooperation: Agent-based Models of Competition and Collaboration. Princeton University Press, Princeton (1997). ISBN: 978-0691015675
13. Bonabeau, E.: Agent-based modeling: methods and techniques for simulating human systems. Proc. Natnl. Acad. Sci. **99**(3), 7280–7287 (2002)
14. Himanshu, K., Ventikos, Y.: Investigating biocomplexity through the agent-based paradigm. Brief. Bioinform. **16**(1), 1–16 (2015)

15. Sharma, V., Compagnoni, A., Libera, M., Muszanska, A.K., Busscher, H.J., Van Der Mei, H.C.: Simulating anti-adhesive and antibacterial bifunctional polymers for surface coating using BioScape. In: Proceedings of the International Conference on Bioinformatics, Computational Biology and Biomedical Informatics (2013)

16. De Giglio, E., Cometa, S., Cioffi, N., Torsi, L., Sabbatini, L.: Analytical investigations of poly (acrylic acid) coatings electrodeposited on titanium-based implants: a versatile approach to biocompatibility enhancement. Anal. Bioanal. Chem. **389**(7–8), 2055–2063 (2007)

17. Wilensky, U., Rand, W.: An Introduction to Agent-Based Modeling: Modeling Natural, Social, and Engineered Complex Systems with NetLogo. MIT Press, London (2015). ISBN: 9780262731898

Models for the Prediction of Antimicrobial Peptides Activity

Rosaura Parisi[1], Ida Moccia[1], Lucia Sessa[1], Luigi Di Biasi[1,2],
Simona Concilio[3], and Stefano Piotto[1(✉)]

[1] Department of Pharmacy, University of Salerno, Via Giovanni Paolo II, 132,
84084 Fisciano, SA, Italy
piotto@unisa.it
[2] Department of Informatics, University of Salerno, Via Giovanni Paolo II, 132,
84084 Fisciano, SA, Italy
[3] Department of Industrial Engineering, University of Salerno,
Via Giovanni Paolo II, 132, 84084 Fisciano, SA, Italy

Abstract. Antimicrobial peptides AMP are small proteins produced by the
innate immune system in multicellular microorganisms. The mechanism of
action of AMP on target membranes can be divided in two main categories: pore
forming and non-pore forming mechanisms. We applied a computational
approach to design novel linear peptides having high specificity and low toxicity
against common pathogens. We built up QSAR models using the data present in
a database of antimicrobial peptides. Here, we present new models of activities
obtained by the use of evolutionary methods and the relative statistical
validation.

1 Introduction

The drug resistance is a limit to the choice of an efficient antibiotic therapy. The reason
is that any microorganisms, through different strategies, can cancel out the action of
antibiotics. Unfortunately, the indiscriminate use of antibiotics accelerated this phenomenon. A classic example of antibiotic resistance is represented by the strain
methicillin resistant *Staphylococcus aureus* (MRSA) [1]. Consequently, there is the
need for new drugs active against pathogens. One of the most promising strategy
against various pathogenic microbes is represented by antimicrobial peptides (AMP).
They are small proteins produced by multicellular organisms that inhibit or kill some
microorganisms (bacteria, fungi, enveloped viruses, protozoans and parasites). AMP
are produced in the innate immune response [2]. These peptides, often small and
cationic, are secreted into the aqueous phase where they are generally in an unfolded
state, but they fold in the proximity of the target membrane [3]. Most antimicrobial
peptides act on the bacterial cell membrane without specific receptors. How AMP kill
bacteria interacting with the cell membrane is not yet completely understood. In fact,
AMP utilize a wide variety of mechanisms, such as altering the membrane equilibrium,
creating pores, disrupting the membrane, altering the membrane fluidity or docking a
protein receptor [4, 5]. Consequently, their membrane interaction and broad activity
spectra are becoming an ideal target to overcome the resistance resulting from bacterial

© Springer International Publishing Switzerland 2016
F. Rossi et al. (Eds.): WIVACE 2015, CCIS 587, pp. 83–91, 2016.
DOI: 10.1007/978-3-319-32695-5_8

mutations [6]. They are classified, according to their secondary structure, into four categories [7]: α-helical, β-sheet peptides, linear extended antibacterial peptides and the loop antibacterial peptides. To date, more than two thousands natural AMP have been isolated and characterized from different sources and several thousands of synthetic variants have been developed. For example, the most studied family of peptides extracted from mammalians is the family of β-defensins. Some researchers developed an approach to identify conserved motifs in these peptides through a computational tool based on hidden Markov models (HMMs) and a basic local alignment search tool [8]. Sequence analysis of these peptides showed low sequence homology [9] precluding the possibility to create easily a model of activity [10]. For this reason, it became important to try different computational approaches for predicting the activity of antibacterial peptides. Several computational studies permitted to develop algorithms to predict antibacterial peptides with a high accuracy. For example, some researchers using Artificial Neural Network (ANN) and Support Vector Machine (SVM) suggested that N- and C-terminals of the AMP sequence might play an important role in the activity: C-terminal is involved in the interaction with the membrane and in the pore formation, while the N-terminal helps in bacteria specific interaction process [10]. The starting point of this work was the selection of sets of homogenous AMP in terms of chemical-physical properties. This step was essential to cluster peptides acting with similar mechanisms. On these sets, we performed a QSAR analysis to determine the relationship between the structural properties of AMP, such as charge, Boman index, or flexibility, with the antimicrobial activity of these molecules (MIC, minimum inhibitory concentration). These sets were analyzed by artificial neural networks and genetic algorithms. In quantitative structure - activity relationships (QSAR) we correlate the biological activity of a class of compounds with the chemical - physical characteristics or structural properties of the compounds themselves. The main limitation of the QSAR studies is the complexity of a biological system. Genetic Algorithms (GA) are heuristic search methods based on the Darwinian theory of natural selection [11]. The artificial neural network (ANN) have been developed and designed to mimic the information processing and learning in the brain of living organisms. The ANN offer satisfactory accuracy in most cases but tend to over fit the training data. Here we present activity models on a gram positive bacterium: *Staphylococcus aureus*.

2 Materials and Methods

The working hypothesis is that peptides with similar features can share the same mechanism of action. We have chosen the parameters present in the database Yadamp [12] to create uniform subsets. We have selected 6 parameters (charge at pH 7, length, CPP index, flexibility, ΔG, helicity as listed in the server Yadamp [12]), and we generated 62 different peptide sets homogeneous in one or two parameters (for example, one set was constituted by the 173 peptides shorter than 30 residues and with a charge at pH 7 between 2 and 7).

On the 62 peptide sets, we applied two kind of mathematical methods.

Genetic algorithms are stochastic optimization techniques that mimic selection in nature that proved to be a very effective tool in QSAR studies. A genetic algorithm chooses a suitable set of descriptors, and the selected descriptors are utilized to build a nonlinear QSAR regression equation. Nonlinear correlations in the data are explicitly dealt with by use of the descriptors in spline, quadratic, offset quadratic, and quadratic spline functions. The method has been implemented in the Material Studio 7.0 [13] package, and it was used here without modification. The smoothness parameter was kept at the default value of 1.0, and the length of an individual was let vary between 2 and 5 descriptors. A total of 500 individuals were let evolve over 5000 new generations.

ANN analysis was performed with the software Matlab 2013 [14]. The multilayers network used have two layers: the output and the hidden layer. The hidden layer consisting of ten artificial neurons, the output layer of a single neuron. The training function of the network is the algorithm based on the Levenberg-Marquardt minimization method (trainlm). This function is very fast and performs better on function fitting (nonlinear regression) problems. The adaption learning function is learngdm, that corresponds to the momentum variant of back propagation. The two different transfer functions used for the neurons are: tan–sigmoid transfer function (tansig) for the hidden layer, that returns values between -1 and 1, and linear transfer function (pureline) for the output layer. The performance function for the network is mean square error (mse).

3 Results

3.1 QSAR Analysis - GA

On each peptide set, we applied the same GA protocol. We identified two equations describing biocidal activity. The R^2 was of 0.92 and 0.81 respectively. Equation 1 was obtained from a dataset of peptides having a length between 7 and 11 amino acids (55 peptides). Equation 2 was obtained using peptides shorter than 30 amino acids and a Boman index between 1 and 2 kcal/mol for a total of 92 peptides. In Eq. 1 the critical parameters for antimicrobial activity are the peptide charge in acid and neutral solution and the number of polar amino acids in the sequence. Equation 2 is similar to Eq. 1 and gives similar importance to peptide charge.

$$MIC = 8.16\,POLAR\,AA - 2571(-0.72 - Ch5)^2 + 9963(-0.90 - Ch7)^2 + 11 \quad (1)$$

$$MIC = -\frac{(MW - 881)^2}{250000} + 122(D - 1.7)^2 + 3134(1.07 - Ch5)^2 - 3340(0.79 - Ch7)^2 + 22$$

$$(2)$$

The parameter function returns the value of the argument, if it is positive, and zero otherwise.

D: Number of residues of Aspartic acid
Ch5: peptide charge at pH5
Ch7: peptide charge at pH7
POLAR AA: number of polar residues
MW: Molecular weight

Both equations confirm that AMP belonging to that set, act through electrostatic interactions with bacterial membrane [15]. However, a good R^2 cannot capture the quality of an activity model because the intrinsic experimental error in microbiological tests, due to serial dilutions, is not considered. It is more correct to talk about activity classes, and the goodness of a QSAR model must be judged in terms of its ability to discriminate among very active, active and non-active peptides. For this reason, MIC (minimum inhibitory concentration expressed in μM) values of 0.3 and 1.8 must be considered as peptides with the same activity. To evaluate the models, we divided the peptides in classes of MIC as shown in Table 1. The 5 classes have similar dimension.

Peptides of classes A, B, C, D are considered active, whereas class E corresponds to inactive peptides.

Table 1. Division of antimicrobial peptides into five classes based on the values of MIC in μmol/mL.

A	B	C	D	E
$0 \leq MIC \leq 2$	$2 < MIC \leq 5$	$5 < MIC \leq 10$	$10 < MIC \leq 30$	$MIC > 30$

The MICs have been calculated for all peptides active against *S. aureus* present in the database. We calculated the precision (PPV), the accuracy (ACC), the sensitivity (TPR) and the specificity (SPC) as defined in Eqs. 3–6.

$$PPV = \frac{TP}{TP + FP} \qquad (3)$$

$$ACC = \frac{TP + TN}{total\ population} \qquad (4)$$

$$TPR = \frac{TP}{TP + FN} \qquad (5)$$

$$SPC = \frac{TN}{TN + FP} \qquad (6)$$

Whereas TP, FP, TN, and FN stand for True positives, False positives, True negatives and False negatives respectively.

The calculation of these indexes requires an arbitrary definition of what is considered *active* and *inactive*. We followed a common view in the pharma industry to consider *inactive* those peptides with a MIC higher than 30 μM. Therefore, active peptides are those belonging to classes A, B, C and D.

In Fig. 1 we plotted the precision, accuracy, sensitivity and specificity for models obtained by GA analysis. For both models, the behavior is acceptable only for three indexes. Specificity (black lines in figure) is the exception, with values that drop to 25 % for Eq. 2 for peptide longer than 40 amino acids. This is not surprising, since the model was obtained from a dataset of shorter peptides.

Low specificity indicates that models displays many false positives. However, a good R^2 and high precision, accuracy and sensitivity, cannot capture the quality of an

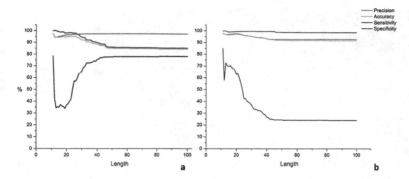

Fig. 1. Evaluation of precision, accuracy, sensitivity and specificity of Eqs. 1(a) and 2(b)

activity model because the intrinsic experimental error in microbiological tests, due to serial dilutions, is not considered. It is more correct to talk about activity classes, and the goodness of a QSAR model must be judged in terms of its ability to discriminate among very active, active and non-active peptides. The overall quality of the model (score) is calculated comparing MIC predictions with the experimental data according to Eq. (7). The scores are indicated in Table 2.

Table 2. Matrix for the computation of the overall model quality

		Observed				
		A	**B**	**C**	**D**	**E**
Predicted	**A**	2	1	0	-1	-2
	B	1	2	1	0	-1
	C	0	1	1	0	-1
	D	-1	0	0	1	0
	E	-2	-1	-1	0	2

$$Score = \sum_{i=1}^{n} Matrix[Class_{observed} - Class_{predicted}] \tag{7}$$

The scoring matrix in Table 2 attributes a reward each time the model correctly predicts the MIC. If the class is not predicted correctly, there is a penalty (negative values). The quality of the model is well represented in Fig. 2. Each point in the figure corresponds to a set of peptides of length between Length_start and Length_stop. The overall quality, calculated with Eq. (7), is rescaled between 0 (blue, unreliable) and 100 (red, reliable), and color mapped.

For example, the point 20, 50 of Fig. 3a indicates that the sum of the scores on all peptides with length between 20 and 50 is lower the 10 %. This diagram permits to easily evaluate the domain of applicability of the model.

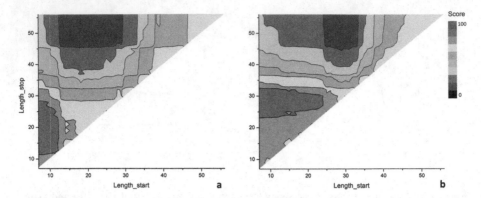

Fig. 2. Results of statistical validation of the Eqs. 1(a) and 2(b) obtained for *S. aureus* (Color figure online)

Figure 2a is relative to Eq. 1. As clearly shown in the diagram, the reliable region (red) is larger than the subset where the model was calculated. For longer peptides, the prediction capability of the model quickly degrade. The Eq. 2 (Fig. 2b) shows a wide reliable region, even larger than the original set of peptides.

3.2 QSAR Analysis – ANN

On the same data sets, we have applied ANN. The neural network used consisted of 2 layers with 10 neurons in the hidden layer. In the first dataset of 55 peptides, the neural

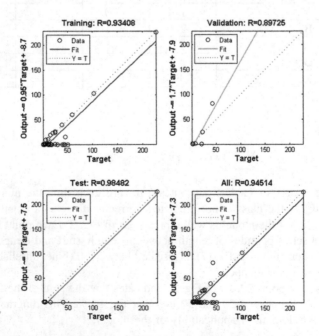

Fig. 3. Results of the application of ANN for peptides with a length between 7 and 11

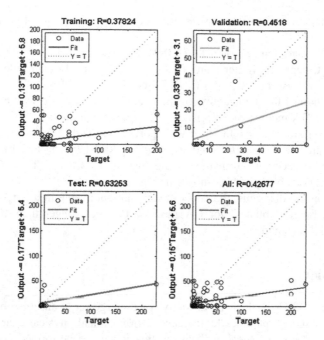

Fig. 4. Results of the application of ANN for peptides shorter than 30 amino acids and a Boman index between 1 and 2 kcal/mol

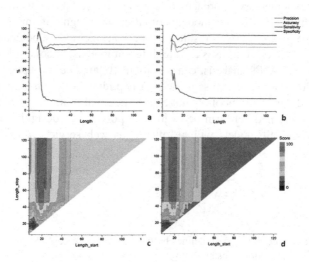

Fig. 5. Result of statistical validation of the two ANN analysis on peptides. The model (a, c) was created from peptides with a length between 7 and 11 amino acids; the model (b, d) was created from peptides shorter than 30 amino acids (b, d)

network found a good correlation between molecular descriptors and the antimicrobial activity.

The overall performance was a R^2 of 0.945, as shown in Fig. 3, whereas on the second data set, peptides shorter than 30 amino acids and a Boman index between 1 and 2 kcal/mol, the overall R^2 was of 0.427 (see Fig. 4).

The evaluation of the applicability of the neural network models were made in the same fashion of GA models. Unsurprisingly, the model is reliable only for the interval between 7 and 11 amino acids. In Fig. 5 we reported the trend of sensitivity, specificity, accuracy and precision for *active* and *inactive* peptides (Fig. 5a and b) for the two models. The more accurate evaluation using the quality matrix (Table 2) assigning peptides to 5 classes of activity is shown in Fig. 5c and d.

As shown in the diagrams, the ANN models are applicable in a range of peptides narrower than ranges obtained for GA models. Peptides longer than 40 cannot be calculated with both models.

4   Conclusion

We conducted a QSAR analysis on the activity of a large set of antimicrobial peptides. The creation of sets of peptides homogeneous in chemical-physical characteristics is indispensable for any statistical analysis. In this work, we performed GA and ANN studies on homogeneous sets of AMP extracted from the peptide database Yadamp. The GA analysis underlined the importance of peptide charge and polarity. This finding support one of most accepted models of activity, that the peptide-membrane interaction is mediated by electrostatic interactions. The artificial neural networks analysis is a complementary approach to GA. We observed a satisfactory fitting of antimicrobial activity only in one model. In that case, though with an $R^2 = 0.945$, the performance score of ANN models resulted lower than GA models, but it can be used for a peptide design based on consensus among different models. In conclusion, the models obtained by GA and ANN analysis, can be efficiently applied to peptides with length between 7 and 20. The number of sequences of peptides shorter than 20, is about 10^{26} that is an extraordinary large pool for novel antimicrobial mining.

The models presented here can be of high importance in designing novel antimicrobial peptides and all models will be offered as web service within the database Yadamp.

References

1. Liu, C., et al.: Clinical practice guidelines by the Infectious Diseases Society of America for the treatment of methicillin-resistant Staphylococcus aureus infections in adults and children. Clin. Infect. Dis. **52**(3), e18–55 (2011) (ciq146)
2. Cruz, J., et al.: Antimicrobial peptides: promising compounds against pathogenic microorganisms. Curr. Med. Chem. **21**(20), 2299–2321 (2014)
3. Cirac, A.D., et al.: The molecular basis for antimicrobial activity of pore-forming cyclic peptides. Biophys. J. **100**(10), 2422–2431 (2011)

4. Török, Z., et al.: Plasma membranes as heat stress sensors: from lipid-controlled molecular switches to therapeutic applications. Biochim. Biophys. Acta (BBA)-Biomembr. **1838**(6), 1594–1618 (2014)
5. Scrima, M., et al.: Structural features of the C8 antiviral peptide in a membrane-mimicking environment. Biochim. Biophys. (BBA)-Biomembr. **1838**(3), 1010–1018 (2014)
6. Marr, A.K., Gooderham, W.J., Hancock, R.E.: Antibacterial peptides for therapeutic use: obstacles and realistic outlook. Curr. Opin. Pharmacol. **6**(5), 468–472 (2006)
7. Wang, G.: Human antimicrobial peptides and proteins. Pharmaceuticals **7**(5), 545–594 (2014)
8. Scheetz, T., et al.: Genomics-based approaches to gene discovery in innate immunity. Immunol. Rev. **190**(1), 137–145 (2002)
9. Hancock, R.E., Chapple, D.S.: Peptide antibiotics. Antimicrob. Agents Chemother. **43**(6), 1317–1323 (1999)
10. Lata, S., Sharma, B., Raghava, G.: Analysis and prediction of antibacterial peptides. BMC Bioinform. **8**(1), 263 (2007)
11. Holland, J.H.: Adaptation in Natural and Artificial Systems: an Introductory Analysis with Applications to Biology, Control, and Artificial Intelligence. MIT Press, Cambridge (1992)
12. Piotto, S.P., et al.: YADAMP: yet another database of antimicrobial peptides. Int. J. Antimicrob. Agents **39**(4), 346–351 (2012)
13. Accelrys, Accelrys Materials Studio. Accelrys Inc., San Diego, California (2014)
14. MATLAB, R.: Version 8.1. 0.604 (R2013a). The MathWorks Inc., Natrick, Massachusetts (2013)
15. Chen, L., et al.: How the antimicrobial peptides kill bacteria: computational physics insights. Commun. Comput. Phys. **11**(3), 709 (2012)

On the Dynamics of Autocatalytic Cycles in Protocell Models

Marco Villani[1(✉)], Alessandro Filisetti[1,2P], Matthieu Nadini[1], and Roberto Serra[1]

[1] Department of Physics, Informatics and Mathematics, University of Modena and Reggio Emilia, v. Campi 213a, 41125 Modena, Italy
{marco.villani, roberto.serra}@unimore.it,
alessandro.filisetti@gmail.com,
matthieu.nadini@gmail.com
[2] Explora s.r.l, Rome, Italy

Abstract. The emergence of autocatalytic sets of molecules seems to have played an important role in the origin of life, allowing a sustainable systems' growth and reproduction. Several frameworks have been proposed, one of the most recent and promising being that of RAF (Reflexively Autocatalytic – Food generated) sets. As it often happens when topological properties only are taken into account, RAFs are however only potentially able of supporting continuous growth. Dynamics can also play a significant role: it is shown here how dynamical interactions may sometimes lead to unexpected behaviors.

1 Introduction

The emergence of autocatalytic sets of molecules seems to have played an important role in the origin of life. Several frameworks have been presented in the past (including those of [1–3]), all sharing the common idea that a core of autocatalytic reactions is needed in order to sustain the system growth and reproduction. One of the recent proposals merges these (Reflexively Autocatalytic) core structures to the chemical pathways needed to guarantee the presence of the material required for the system's growth (the Food) into the so-called RAF (Reflexively Autocatalytic – Food generated) sets [4, 5]. In spite of major differences among these models, a common feature is the presence of a transition so that, when the diversity of the involved molecular types present in the reaction volume exceeds a threshold, catalytic sets do appear.

However, although the possibility to reproduce the emergence of such structures in laboratory has received considerable attention, this is still far from being achieved: the few known examples have been obtained only by carefully engineering the involved types [6], and not through the interaction of random sets of molecules.

One of the main missing parts, possibly able to connect theory and experiments, is the dynamics: the emergence and maintenance of these self-sustaining assemblies could derive not only from some peculiar feature of their topological structures (as the network of reactions linking the different chemical species), but also from the nature of the dynamical processes the interaction between their structures and the environment

F. Rossi et al. (Eds.): WIVACE 2015, CCIS 587, pp. 92–105, 2016.
DOI: 10.1007/978-3-319-32695-5_9

can allow. In other words, dynamics can allow – or inhibit – the unfolding of the self-organizing properties of the identified structures.

In particular, the dynamics of all current living beings happens in small confined structures that embed and confine catalytic sets of reactions thus leading to their growth and reproduction. A part of the literature models these structures by using the approximation of the CSTR (Continuous flow Stirred-Tank Reactor) framework [7–9], but the elementary units of living beings are very different from these simple chemical reactors [5, 10]. Unlike CSTRs, where the nutrients are externally determined, in cells the exchange of matter and energy strongly depends on their membrane properties and on their internal chemical composition.

There are several different proposals about the structure of protocells, cell-like structures - simpler than present-day biological cells - but nonetheless able (i) to grow by some form of rudimentary metabolism, (ii) to reproduce giving rise to new proto-cells that are similar to their parents and (iii) to undergo evolution [7–9, 11–22]. The protocell study could be important in order to create new "protolife" forms which are able to adapt and evolve [7, 15].

So in this article we investigate the dynamical behavior of RAF sets within pro-tocells, finding interesting and unexpected behaviors. Section 2 presents the ideas and the models, with the techniques needed to represent and examine the chemical reaction system; Sect. 3 presents the performed experiments and their results; Sect. 4 highlights the main conclusions.

2 The Model

2.1 The Chemistry

The world that the protocells inhabit is defined by a set of chemical species, reactions and catalyzes (a "chemistry" in the following); in the following we use several "random generated" chemistries,[1] which share (or differ in) characteristics as number of species and number of reactions. The species of each chemistry are composed by linear polymers, as described in [2, 8, 9, 23].[2]

The basic entities of these systems are monomers and polymers, identified by ordered strings of letters taken from a finite alphabet (e.g. A, B, C,...). In the following we often refer to these "letters" with the term "monomers", being clear from the context if these objects are constituting the "bricks" of a polymer or independent molecular types composed of a single brick.

Let $X = \{x_1, x_2, \ldots x_N\}$ the whole set of N chemical species composing the chemistry (x_i being the amount – the number of molecules - of the species x_i). The two basic reactions are (i) condensation, (the concatenation of two species in a longer one - e.g. BB + AAB→BBAAB) and (ii) cleavage (the cutting of a species composed of more than one part into two shorter species – BBBA→B + BBA).

---

[1] The usual basic choice for network-based approaches – see also [2, 9].

[2] Note that the aim of the model is not to provide a detailed description of a specific set of reactions; rather, it wants to focus the attention on the general characteristics emerging from the interaction of a large number of interacting molecules.

In this paper we consider that the rates of the spontaneous reactions are very low with respect to that of the catalyzed ones: so, any reaction occurs only if catalyzed by one of its specific catalysts. In order to avoid unrealistic collisions among three different objects, condensations require an intermediate reaction in which a temporary complex between the catalyst and the substrate is formed. By supposing that the example just shown be catalyzed by chemical species BABA we can write:

(a) BABA + BB→ *BABABB (first condensation step)
(b) *BABABB→ BABA + BB (spontaneous complex dissociation)
(c) *BABABB +AAB→BBAAB + BABA (second condensation step releasing the catalyst and the final product)

the "*" of *BABABB indicating that this string is representing a temporary complex and not a normal chemical species.

Not all the chemical species can exhibit catalytic activities; in the following simulations we suppose that only species constituted by at least L_{mincat} bricks can be catalysts[3], with a given probability p of catalyzing any reaction;[4] each chemistry potentially involves all species till length L_{max}.[5]

Finally we neglect the presence of backward reactions (exception made for the dissociation reaction of intermediate complexes) by hypothesizing that the Gibbs energy for any reaction is sufficiently large[6].

As anticipated, following [2] and [23] we create each chemistry randomly; a condensation is defined by randomly choosing two substrates (the first being involved with the catalyst on the complex formation) and a cleavage by randomly choosing one substrate and a cutting point. All the chemistries used during this work divide their reactions in about 50 % of condensation and 50 % of cleavages. Any species x_i (in the following simulations composed by the binary alphabet {A,B}) having length greater than L_{mincat} has a finite fixed probability p_i of being chosen as catalyst of a randomly chosen reaction, leading to the average catalysis level (or also average "connectivity") $<c>$ = (number_of_catalysis)/(number_of_chemical_species). So, each chemistry can be described by the vector {L_{mincat}, L_{max}, $<c>$}, the number of the involved chemical species being $N = {}^{2L_{max}+1}-2$.

---

[3] It is generally believed that L_{mincat} should not be very small (chemical species should reach an enough high complexity to carry catalytic activities) [24]. However, there could be significant exceptions [25].

[4] The probability p is fixed, and the basic model does not hypothesize particular functional relationship between sequences of the catalysts and the reactions they catalyze, as for example chemical affinities among molecules because of their internal composition. These limits do not affect the description capabilities of the model, as discussed in [7].

[5] Note that L_{mincat} and L_{max} help in defining some characteristics of the simulated artificial word, without changing in significant way its dynamics. A different role is played by p, as we discuss in the following.

[6] We deal with the backward reactions theme in [26], and with the energy theme in [27].

2.2   Autocatalytic Structures

Viable protocells require that the internal materials are able to growth at the same rate of the container. In the literature there are several proposals of structures showing this characteristic; in this paper we focus our attention on the RAF sets, groups of reactions guaranteeing the production of all the needed chemical species (starting from a small set of chemical that is guaranteed by the environment) [4].

More precisely, a RAF set is defined as a subset $R' \subseteq R$ of all possible reactions (and associated molecule types) which is:

1. reflexively autocatalytic (RA): each reaction $r \in R'$ is catalyzed by at least one molecule type involved in R'
2. food-generated (F): all reactants in R' can be created from the food set F by using a series of reactions only from R' itself

A more mathematical definition is provided in [4], including an efficient algorithm for finding RAF sets in a general catalytic reaction set. RAFS can be represented by means of the graph substrates/products-reactions-catalysis, often simplified by the catalyst-product representation. In this last representation autocatalytic structures acquire the form of SCCs (Strongly Connected Component), structures where each node is directly or indirectly reachable starting from any other node of the same structure: this partial picture however does not allow the recognition of the substrates needed for the system's growth (see Fig. 1 for a more detailed presentation).[7]

Indeed, as many other characterizations of interesting groups based on topological aspects of some graph representation, RAFs are only *potentially* able of supporting continuous growth: as far as we know no papers in the literature focus their attention on the dynamical behavior of RAF sets in dynamically plausible protocell systems. This paper shows counterintuitive behaviors and better characterizes interesting aspects of the dynamical behavior of RAF sets in protocells.

2.3   The Protocell Architecture

In this paper we consider protocells composed by a finite volume enclosed by a membrane that - for simplicity – is assumed to be permeable only to short molecules. The transmembrane motion of the permeable species is here supposed to be ruled by the difference of their chemical potentials in the internal aqueous volume of the protocell and in the external aqueous environment. We assume that transmembrane diffusion is extremely fast, so that there is always equilibrium between the internal and external concentrations[8] of the permeable species (this adiabatic hypothesis could be

---

[7] Moreover, RAFs can be composed also by linear chains of reactions, provided that the presence of the first element of the chain is guaranteed by the environment (a case not present in this work, because of the chemical species that can pass through the membrane have not catalytic activities).

[8] Given our assumptions the difference in chemical potential is due only to differences in concentrations.

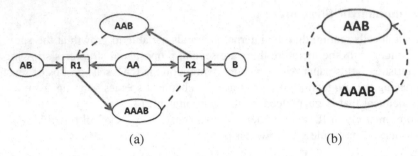

Fig. 1. The figures shows a particular RAF set on (a) the complete graph substrates/products-reactions-catalysis and on (b) the simplified catalyst-product graph representation. In (a) solid lines represent materials production/consumption, whereas dotted lines represent catalysis; AA, AB and B species are needed for the production of species AAB and AAAB and are provided by the environment. Scheme in (b) represents the same process, that in this case constitutes a SCC (Strongly Connected Component– a structure where each node is directly or indirectly reachable starting from any other node of the same structure); this partial representation does not allow the recognition that some materials have to be present to allow the system's growth

easily relaxed in the future). Moreover, the external volume is supposed very large with respect to the total volume of the present protocells – so, the concentration of these molecules in the internal volume of the protocell could be considered constant (note that in such a way their flow through the membrane depends on their participation in the internal chemical dynamics). We assume that only short molecules can pass through the membrane – in the following, all molecules shorter than L_{buff} monomers. For simplicity, we assume also that $L_{buff} = L_{mincat}$.

In our model (a subset of) the chemical species belonging to a RAF can catalyze the membrane growth [5, 12]:

$$\frac{dC}{dt} = \sum_{i=2^{L_{buff}-2}}^{N} \alpha_i[x_i]V_r \qquad (1)$$

where V_r is the internal volume of the protocell (where reactions occur) and $[x_i]$ is the concentration of catalysts in the internal aqueous phase; the kinetic coefficients α_i are zero for all those species that do not contribute to the container growth.

The membrane at a given threshold splits into two daughter protocells, each owning half of the membrane materials and about the 36 % of the internal materials [5]. With a different model we already demonstrated that – under certain conditions – the growth rate of membrane and internal materials synchronizes reaching a sustainable collective growth[9] [12, 28], but no mention was made of RAF sets. In this paper how their presence and their dynamics can affect the overall properties of protocells.[10]

3 Results

3.1 Different Chemistries

An interesting hypothesis is that the higher is the number of involved chemical reactions, the higher is the probability that the obtained chemical soup can support self-organizing structures [2, 32]. Figure 2a shows that the probability that a randomly created chemistry contains SCC or RAF structures grows with the increase of the average connectivity <c>. Both SCCs and RAFs show a sudden and significant increase (a sort of phase transition) in some zones; the fact that these zones significantly differ from each other could be indicated as one of the reasons that at present are preventing the observation of the emerging of autocatalytic structures in wet laboratories (that is, experimentalists are working close the SCC's critical point rather than near the RAF's critical point, not providing in such a way to reactions enough materials to proceed) [5].

It is possible however to select chemistries close to both critical points that have potentially working RAFs: the interesting question is whether and how these RAFs are effectively working in protocell-like structures, dynamical frameworks able to expand and divide.

So we selected 20 chemistries owning a RAF at <c> = 1.0 (the connectivity assuring 50 % of probability of finding a SCC on a random chemistry) and 20 chemistries owning a RAF at <c> = 2.5 (the connectivity assuring 50 % of probability of finding a RAF on a random chemistry), both groups having $L_{mincat} = 3$ and $L_{max} = 6$ (so, $N = 126$). We remember that in order to avoid banal behaviors we impose that the chemical species passing through the semipermeable membrane are not able to catalyze reactions (the $L_{buff} = L_{mincat}$ condition): as a consequence the found RAFs are necessarily composed by SCCs[11]. Finally, we performed the simulations by coupling the

---

[9] Note that the other possible outcomes are dilution (at division the internal materials have always smaller concentrations, leading to huge division times and starvation) and excessive concentration (at division at least a part of the internal materials have always higher concentrations, leading to the protocell breakage). This last event is out of the model's range of validity, but can be simply detected by observing the internal material concentrations and stopping the simulation in case of too high concentration values. In the following we refer to both these events as "not synchronizing situations".

[10] The great part of the simulations presented in this work was performed by using a home-made tool implementing stochastic dynamics (an extension of the Gillespie algorithm [29] to the case of variable volumes, already presented in [30]), because in some cases the effects of randomness may be very relevant [10]. When concentrations were sufficiently high the simulations was double-checked by means another tool implementing a deterministic Euler schema using a with step size control (already used in [31]).

[11] In order to find 20 different chemistries satisfying this last vinculum we created 600 different chemistries having <c> = 1.0 (580 chemistries discarded) and 50 different chemistries having <c> = 1.0 (30 chemistries discarded).

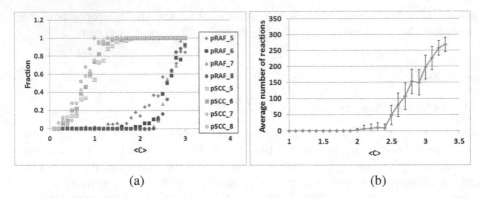

(a)   (b)

Fig. 2. (a) The fraction of simulations showing at least 1 RAF (left) and 1 SCC (right), by varying <c> and M (from 5 to 8) in networks with forward reactions only. F contains all the species up to length 2. On the x-axis the average level of catalysis <c> is represented while on the y-axis the fraction of network instances (out of 100 networks for each <c>) is depicted. (b) The average number of reactions involved on RAF for the chemistries with $M = 6$ (out of 100 networks for each <c>)

chemical species belonging to the RAF with the membrane; in particular we used three different levels of interaction, in each group all coupling coefficient of the involved species having the same value (respectively, $\alpha = 0.1$, $\alpha = 0.01$ and $\alpha = 0.001$, α referring to Eq. 1).

Unexpectedly, despite the fact that RAFs at <c> = 1.0 are significantly smaller than RAFs at <c> = 2.5 (these groups being respectively composed by RAFs including 2−3 reactions instead of RAFs owning typically several tents of reactions – se also Fig. 2b), the biggest part of these RAFs are able to support the protocell growth, in contrast with the only 10 % (or less) of RAFs at <c> = 2.5.

A second interesting finding of this research indicates that the lower is the coupling coefficient value, the higher is the probability of having synchronization (see Table 1).

Finally, the two sustainable protocell growths in chemistries at <c> = 2.5 are supported respectively by only 2 chemical species (over the 2 species included on the RAF – by far the smallest RAF of this group of chemistries) and 3 species (over the 42

Table 1. Number and percentage of synchronizing protocells at different level of coupling between RAF's chemical species and membrane. Note that chemistries synchronizing at $\alpha = 0.1$ synchronize also at $\alpha = 0.01$, and chemistries synchronizing at $\alpha = 0.01$ synchronize also at $\alpha = 0.001$ (a situation not directly indicated in table)

Chemistries with <c>= 1.0			Chemistries with <c> = 1.0		
α	Number of synchronizing chemistries	Percentage of synchronizing chemistries	α	Number of synchronizing chemistries	Percentage of synchronizing chemistries
0.1	12	60 %	0.1	1	5 %
0.01	19	95 %	0.01	2	10 %
0.001	19	95 %	0.001	2	10 %

species included on the RAF). So, there should be some process or situation hampering the deployment of the potentialities of great size RAFs. We call the set of species included in the RAF and enabling a sustainable protocell growth sRAF (synchronizing RAF); in the following sections we investigate some properties of these groups.

3.2   The Inner Structure of SRAFs

In random chemistries big size RAFs do not support the protocell growth. Some hints about the reasons of this limitation can be deduced from the observation of sRAFs supporting the protocell synchronization at low α values but not at the high ones: all these sRAFs share the property of using a part of their species as reagents in order to build other species of the sRAF itself (see Fig. 3b and c for some examples). Interestingly such a kind of RAF was already identified in [7], but was at the same time was classified as "suicidal RAF", not suitable for supporting useful functions in living structures. On the contrary, we find that in presence of a non-excessive coupling with the membrane this kind of structures can effectively sustain the protocell growth.

At the same time an excessive number of suicidal parts can highly reduce the reproducing capabilities of a RAF structure, and RAFs having big dimensions have a high probability of owning many of these suicidal parts. So, RAF belonging to random chemistries can hardly be involved in sustainable protocell growth.

Another observation allows a similar although slightly subtle consideration: no cleavages are present in the observed sRAFs. We remember that in our model only short chemical species can cross the membrane, whereas cleavages necessarily have as substrates relatively long species, that in turn need an inner active production in order to maintain their presence along the protocells' generations. So, as we can observe in Fig. 3c, the sRAF collectively destroys a chemical species that in turn have to rebuild in order to allow again the production of his other parts. It is a "collectively suicidal" behavior, very similar to that of the just presented one – so, its low probability of being included in sRAF.

Interestingly, the combination of the relatively short length of the membrane crossing chemical species and of the weakening influence of the "collectively suicidal" processes (and in particular of cleavages) could introduce in protocells a symmetry breaking, that forces protocells to build long rather than short molecules. We are expanding this point in next sections.

3.3   Competition Among SRAFs

Different sRAFs contemporaneously present inside the same protocell can compete through their actions on the membrane; in this case different outcomes are possible, the most probable situation being the surviving of only one sRAF. Figure 3 shows different sets of reactions and chemical species, each group being able to synchronize with the membrane growth if present alone within a protocell (each one constituting therefore a sRAF).

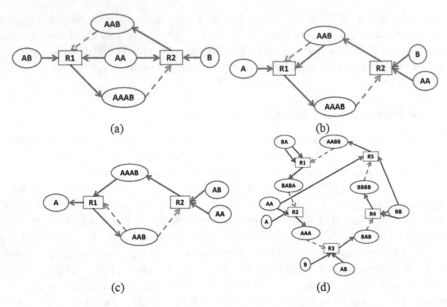

(a)   (b)

(c)   (d)

Fig. 3. The structure of four sRAFs used in this section. (a) a sRAF whose substrates can cross through the membrane and therefore are continuously provided by the environment (sRAF_A); (b) a sRAF where one of the reactions uses as substrate its catalyzer (a "suicidal process") (sRAF_B); (c) a SRAF composed by one condensation and one cleavage (a "collectively suicidal" process where we assist to the continuous creation (reaction R2) and destruction (reaction R1) of species AAAB – in this case both actions being catalyzed by the same catalyst AAB) (sRAF_C); (d) a SRAF composed by 5 reactions (all condensations) whose substrates are continuously provided by the environment (sRAF_D). Solid lines represent materials production/consumption, whereas dotted lines represent catalysis; if not differently indicated on the text, all the kinetic constants of the reactions have the same values

If not differently indicated in the following, we consider all the kinetic constants of the reactions as having the same values[12]: if this situation do not occur indeed the chemical groups connected with the higher growth rate force a protocell growth and duplication to so high rates that during generations the other chemical species dilute and disappear (a different behavior being possible only by changing some model features, as discussed on the conclusions). In any case, ceteris paribus, different organizations of chemicals and reactions can still lead to different outcomes.[13]

---

[12] In particular, the kinetic Gillespie constant for the first step of the condensations (the union of the first substrate with the catalyst to build a short-life complex) is $cs = 8.3e^{-3}$, the constant for the first-order complex dissociations is $c_s = 250$, the constant for the final step of the condensations (the union of the second substrate with the complex releasing the catalyzer and the final product)) is $c_s = 8.3e^{-4}$, and finally the constant for the cleavages is $c_s = 8.3e^{-5}$.

[13] Note that the particular subdivision chosen for the condensation process involves steps having different kinetics: in particular, steps (1) and (3) described in Sect. 2.1 involve collisions between two objects, whereas step (2) follows a simple first order kinetic. So, if the volume of the reaction container changes in time (as in our case) the relative reaction rates can correspondingly change: the effect of this phenomenon on the competition among sRAF will be explored in next works.

As anticipated in previous section, when embedded on the same protocell sRAF_A (composed only by condensations) dilute sRAF_B (where a suicidal loop appears) at any α value. A similar outcome happens when we use sRAF_A and sRAF_C (where is present a cleavage); in this case however at very low α values ($\alpha = 0.0001$) both can coexist, the species belonging to sRAF_C maintaining at protocell duplication a concentration lower than the species of sRAF_A by several order of magnitude – so, the directly suicidal loop is a combination more weakening than the presence of a cleavage (as its dilution when in competition with sRAF_C demonstrates).

sRAF_A and sRAF_B (two structures having the same "building blocks" but being composed by a different number of species and reactions) can coexist; however, if only one kinetic constant is different the group corresponding to the lower growth rate dilute and disappear. Interestingly, at low concentrations the fluctuations induced by the system stochasticity affect more incisively the smaller structure, that have therefore higher chances of disappearing: in long runs the surviving protocells are composed mostly by the bigger sRAF. This symmetry breaking is nevertheless weak, and systems having sRAFs respectively composed by 5 and 10 chemical species are enough robust to make probable their contemporaneously surviving (this contemporaneous surviving having however lower probability in case of high coupling with the membrane).

3.4 Varying the Coupling with the Membrane

Interestingly, in all our simulations (i) the protocell duplication times are very near each other's independently from the coupling coefficient α and (ii) the product between duplication time and total concentration of sRAF species at duplication is inversely proportional to the coupling coefficient α, that is:

$$T_d C_f = \frac{K}{\alpha} \qquad (2)$$

These results can be understood by using simplified deterministic models, as that presented in [12]. By taking into account a very simple RAF, composed by the only chemical species X, and supposing buffered the needed substrates a protocell can be described by the system:

$$\begin{cases} \dfrac{dC}{dt} = \alpha X \\ \dfrac{dX}{dt} = \eta X \end{cases} \qquad (3)$$

In [33] we proved that:

$$T_d = \frac{\ln(2)}{\eta} \qquad (4)$$

$$D_f = \frac{\theta \eta}{2\alpha} \qquad (5)$$

where θ and D_f indicate respectively the quantity of lipids and the quantity of X at duplication time. By multiplying (4) and (5) we obtain:

$$T_d D_f = \frac{\ln(2)\theta}{2\alpha} \approx \frac{K}{\alpha} \qquad (6)$$

which is proportional to α^{-1} (for each protocell θ - and so the protocell's volume - being a constant quantity). In most our simulations we found $T_d \approx 110s$. and $K \approx 0.0005$. Remarkably, the only exceptions are a couple of simulations at intermediate α level that synchronize with $T_d \approx 600s.$, when the same sRAF at $\alpha = 0.1$ do not synchronize and at $\alpha = 0.001$ "regularly" synchronize at 110 s. Our interpretation is that in these cases the coefficient α is very near to the critical situation dividing synchronizing and not synchronizing protocells: in these cases protocells shows a significant reduction of performances – and stochasticity can likely play a significant role.

Finally, some chemistries show an interesting behavior by varying the α coupling coefficient. Whereas the relative concentration values of the species produced by sRAF that use as substrates the simple chemicals provided by the environment are not affected by coupling coefficient α, for different situation (similar to the "suicidal" arrangements of sRAF_B and sRAF_C) the same quantities are different at different coupling coefficient α.

Figure 4 shows the sRAF of chemistry$14_{2.5}$ (the fourteenth of the 20 chemistries having $<c> = 2.5$) and the final quantities of its chemical species (averaged over the last 10 generations), whereas Fig. 5 shows their behavior through the 50 observed generations. It is possible to observe the absence of synchronization at $\alpha = 0.1$ and the change of the concentration rank between species BBA and BBAB by changing α from 0.01 to 0.001.

Chemistry$14_{2.5}$ and the other chemistries showing a similar behavior share the property of having at least one reaction that uses as substrate one species belonging to the sRAF, whereas all other reactions uses only substrates directly furnished by the

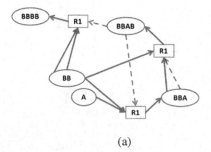

α	BBA	BBAB	BBBB
0.1	0	0	0
0.01	68	17	87
0.001	1091	1316	2449

(a)   (b)

Fig. 4. (a) The "suicidal" sRAF of chemistry$14_{2.5}$ (the fourteenth of the 20 chemistries having $<c> = 2.5$) and (b) the final quantities (in number of molecules, averaged over the last 10 generations) of its chemical species

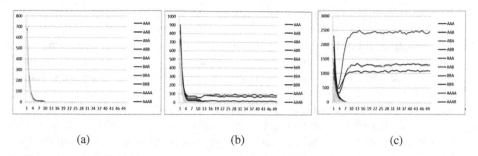

(a)   (b)   (c)

Fig. 5. Time behavior (protocells' generations) of the protocell owning the suicidal sRAF of chemistry$14_{2.5}$, at the coupling coefficients (a) $\alpha = 0.1$, (b) $\alpha = 0.01$ and (c) $\alpha = 0.001$: only the quantities (number of molecules) of the chemical species at duplication time are shown. It is possible to observe the dilution of the chemicals during 50 generations: only the quantities of the species composing the sRAF are asymptotically different from zero.

environment. So, all these reactions have always the same relative rates – independently from α - whereas the "suicidal" part rate depends on the concentration of a substance whose concentration varies with α (as demonstrated by Eq. (6)) – so the changing of the concentration's ranks.

4 Conclusions

In the literature many frameworks potentially able to explain the emergence of autocatalytic sets of molecules are described; in all these frameworks the formation of these assemblies is highly probable, provided that an high enough molecular diversity is present. Nevertheless, it seems that finding these structures in the lab is extremely difficult.

There are many plausible explanations of this distance between theories and experiments, one of the main missing parts regarding the dynamics: the emergence and maintenance of self-sustaining groups could derive not only from some peculiar feature of their topological structures (a point stressed by many researchers), but also from the nature of the dynamical processes the interaction between their structures and the environment allow. So, dynamics can allow – or inhibit – the unfolding of the self-organizing properties of the identified structures.

In this paper we explore the dynamical behavior of one of the most interesting autocatalytic structure, the so-called RAF, that links a (Reflexively Autocatalytic) core with the chemical pathways building the needed substrates (the Food), when embedded in a protocell architecture.

Interestingly, our analysis shows that:

- (unexpectedly) chemistries owning a RAF composed by a high number of chemicals often do not favor the emergence of sustainable protocell growth;

- the lower is the influence of the RAF species on the growth rate of the membrane, the higher is the probability of a successful synchronization between the growth rates of the membrane and of the internal materials, although times may become long;
- a RAF is composed by several independent structures; in our model typically only one of these parts drives the asymptotic protocell growth (the so called synchronizing RAF, or sRAF), whereas the others dilute;
- ceteris paribus, sRAFs that uses as substrates a part of their own components (i) in absence of other sRAFs can sustain the protocell growth, whereas (ii) in presence of other kinds of sRAF they will dilute. The combination of this feature with the fact that typically only a subset of the chemical species can pass through the membrane can induce a symmetry breaking, that forces protocells to build long rather than short molecules;
- there is a mathematical relationship among the value of the internal concentration, the asymptotic duplication time and the coupling strength with the membrane.

So, the coupling of RAF and membrane could sustain a very interesting dynamics where different kinds of protocells could emerge, change and compete, providing the basis for the beginning of an interesting evolutionary dynamics.

Acknowledgments. Useful discussions with Stuart Kauffman, Timoteo Carletti, Chiara Damiani, Alex Graudenzi, Wim Hordijk and Irene Poli are gratefully acknowledged.

References

1. Eigen, M., Schuster, P.: The Hypercycle - A Principle of Natural Self-Organization. Springer-Verlag, Berlin (1979)
2. Kauffman, S.A.: Autocatalytic sets of proteins. J. Theor. Biol. **119**, 1–24 (1986)
3. Jain, S., Krishna, S.: A model for the emergence of cooperation, interdependence, and structure in evolving networks. PNAS **98**(2), 543–547 (2001)
4. Hordijk, W., Steel, M.: Detecting autocatalytic, self-sustaining sets in chemical reaction systems. J. Theor. Biol. **227**, 451–461 (2004)
5. Villani, M., Filisetti, A., Graudenzi, A., Damiani, C., Carletti, T., Serra, R.: Growth and division in a dynamic protocell model. Life **4**, 837–864 (2014)
6. Vaidya, N., Manapat, M.L., Chen, I.A., Xulvi-Brunet, R., Hayden, E.J., Lehman, N.: Spontaneous network formation among cooperative RNA replicators. Nature **491**, 72–77 (2012)
7. Vasas, V., Fernando, C., Santos, M., Kauffman, S.A., Szathmary, E.: Evolution before genes. Biol. Direct **7**, 217–222 (2012)
8. Filisetti, A., Graudenzi, A., Serra, R., Villani, M., De Lucrezia, D., Füchslin, R.M., Kauffman, S.A., Packard, N., Poli, I.: A stochastic model of the emergence of autocatalytic cycles. J. Syst. Chem. **2**, 2 (2011)
9. Filisetti, A., Graudenzi, A., Serra, R., Villani, M., Füchslin, R.M., Packard, N., Kauffman, S.A., Poli, I.: A stochastic model of autocatalytic reaction networks. Theor. Biosci **131**(2), 85–93 (2012)
10. Serra, R., Filisetti, A., Villani, M., Graudenzi, A., Damiani, C., Panini, T.: A stochastic model of catalytic reaction networks in protocells. Nat. Comput. **13**, 367–377 (2014)

11. Serra, R., Carletti, T., Poli, I.: Synchronization phenomena in surface-reaction models of protocells. Artif. Life **13**, 123–138 (2007)
12. Carletti, T., Serra, R., Villani, M., Poli, I., Filisetti, A.: Sufficient conditions for emergent synchronization in protocell models. J. Theor. Biol. **254**, 741–751 (2008)
13. Mansy, S.S., Schrum, J.P., Krishnamurthy, M., Tobé, S., Treco, D.A., Szostak, J.W.: Template-directed synthesis of a genetic polymer in a model protocell. Nature **454**, 122–125 (2008)
14. Morowitz, H.J., Heinz, B., Deamer, D.W.: The chemical logic of a minimum protocell. Orig. Life Evol. Biosph. **18**, 281–287 (1988)
15. Szostak, J.W., Bartel, D.P., Luisi, P.L.: Synthesizing life. Nature **409**, 387–390 (2001)
16. Munteanu, A., Solé, R.V.: Phenotypic diversity and chaos in a minimal cell model. J. Theor. Biol. **240**, 434–442 (2006)
17. Segrè, D., Lancet, D.: Composing life. EMBO Rep. **1**, 217–222 (2000)
18. Ganti, T.: Chemoton Theory, 3rd edn. Kluwer Academic, New York (2003)
19. Rasmussen, S., Chen, L., Nilsson, M., Abe, S.: Bridging nonliving and living matter. Artif. Life **9**, 269–316 (2003)
20. Rasmussen, S., Chen, L., Deamer, D., Krakauer, D.C., Packard, N.H., Stadler, P.F., Bedau, M.A.: Transitions from nonliving to living matter. Science **303**, 963–965 (2004)
21. Shirt-Ediss, B., Ruiz-Mirazo, K., Mavelli, F., Solé, R.V.: Modelling lipid competition dynamics in heterogeneous protocell populations. Sci Rep. **4**, 5675 (2014)
22. Mavelli, F., Luisi, P.L.: Autopoietic self-reproducing vesicles: a simplified kinetic model. J. Phys. Chem. **100**(41), 16600–16607 (1996)
23. Bagley, R.J., Farmer, J.D.: Spontaneous emergence of a metabolism. Artificial life II Santa Fe Institute Studies in the sciences of complexity **10**, 93–141 (1992)
24. Alberts, B., Johnson, A., Lewis, J., Raff, M., Roberts, K., Walter, P.: Molecular Biology of the Cell. Garland Science, New York (2002)
25. Gorlero, M., Wieczorek, R., Adamala, K., Giorgi, A., Schininà, M.E., Stano, P., Luisi, P.L.: Ser-His catalyses the formation of peptides and PNAs. FEBS Lett. **583**(1), 153–156 (2009)
26. Filisetti, A., Graudenzi, A., Damiani, C., Villani, M., Serra, R.: The role of backward reactions in a stochastic model of catalytic reaction networks. In: Liò, P., Miglino, O., Nicosia, G., Nolfi, S., Pavone, M. (eds.) Advances in Artificial Life, ECAL 2013, pp. 793–801. MIT Press, Cambridge (2013)
27. Filisetti, A., Graudenzi, A., Serra, R., Villani, M., De Lucrezia, D., Poli, I.: The role of energy in a stochastic model of the emergence of autocatalytic sets. In: Bersini, H., Bourgine, P., Dorigo, M., Doursat, R. (eds.) Advances in Artificial Life ECAL 2011, pp. 227–234. MIT Press, Cambridge (2011)
28. Filisetti, A., Serra, R., Carletti, T., Villani, M., Poli, I.: Non-linear protocell models: synchronization and chaos. EPJB **77**(2), 249–256 (2010)
29. Gillespie, D.T.: A general method for numerically simulating the stochastic time evolution of coupled chemical reactions. J. Comput. Phys. **22**, 403–434 (1976)
30. Carletti, T., Filisetti, A.: The stochastic evolution of a protocell: the Gillespie algorithm in a dynamically varying volume. Comput. Math. Methods Med. **2012**, 12 (2012)
31. Serra R. Villani M.: Mechanism for the formation of density gradients through semipermeable membranes. Phys. Rev. E **87**, 062814 (2013)
32. Kauffman, S.A.: The Origins of Order. Oxford University Press, UK (1993)
33. Serra, R., Carletti, T., Poli, I.: Synchronization phenomena in surface-reaction models of protocells. Artif. Life Spring **13**(2), 123–138 (2007)

A Combined Preprocessing Method for Retinal Vessel Detection Towards Proliferative Diabetic Retinopathy Screening

Leonarda Carnimeo[✉], Annamaria Roberta Altomare, and Rosamaria Nitti

Department of Electrical and Information Engineering, Technical University of Bari, Bari, Italy
leonarda.carnimeo@poliba.it

Abstract. In this paper, the problem of detecting blood vessels in retinal images with early proliferative retinopathy is faced by highlighting vessels both in retina background and in the optic disc. For this purpose, a Combined Method for preprocessing fundus oculi images is developed. In detail, each retinal vessel image is enhanced via a contrast-limited adaptive histogram equalization after applying a suitable operator for feature extraction. Then, the proposed Combined Preprocessing Method is synthesized to modify each contrast-enhanced retinal image using both a Two-dimensional Matched filtering and a 2D Gabor Wavelet Transformation for vessel highlighting. Combination and segmentation of preprocessed images are subsequently performed and binary maps of retinal vessels are finally derived. The effectiveness of the proposed method is evaluated on obtained outcomes and results are compared to those obtained with available techniques.

Keywords: Retinal Vessel Detection · Preprocessing methods · Medical image classification · Diabetic retinopathy · 2D Gabor wavelets · Matched filters

1 Introduction

In the last years, several improvements have been achieved in medical imaging [1–5]. If ophthalmic advances are considered, at the moment automated diagnostic tools allow clinicians to perform retinal examinations of mass screening for the most common diseases, such as diabetes, hypertension and glaucoma [6–12]. Specific clinical markers help ophthalmologists in diagnosing Diabetic Retinopathy (DR) [7–9]. In this regard the extraction of retinal vessels is essential in the analysis of digital fundus images since it helps in diagnosing several retinal diseases. More in detail, the clinical aim towards the segmentation of blood vessels of retinal images deals with the suppression of the background and the enhancement of all small vessels, so that abnormal neovascularization can become more visually highlighted. Even if the local contrast can be very low in a retinal image, in [10, 11] a segmentation algorithm, which approximates the intensity profile of retinal vessels using a Gaussian curve, was proposed to detect both vessels and other brighter objects, such as lesions, optic disc, etc. Furthermore, among retina diseases, the early diagnosis of proliferative DR, a severe complication of diabetes that damages retina, is crucial to the protection of the vision of diabetic patients [13–16].

© Springer International Publishing Switzerland 2016
F. Rossi et al. (Eds.): WIVACE 2015, CCIS 587, pp. 106–116, 2016.
DOI: 10.1007/978-3-319-32695-5_10

The onset of this disease is signaled by the appearance of neovascular sprouts, which might be identified using some retinal vessel extraction techniques as shown in [13, 14]. In particular, in [14] a procedure of extraction of small retinal vessels is performed, but the main drawback of the approach is that for each considered set of images, all necessary parameters have to be manually set to obtain satisfactory results. In this regard, the use of high values of a particular limiting parameter makes the noise grow in the image, thus increasing the probability of obtaining false detections [14]. Furthermore, beside several edge detection algorithms are available in current literature [1], they do not always lead to acceptable results in extracting various features in a fundus image with proliferative retinopathy. Some drawbacks have not been overcome yet, since every pre-processing procedure provides improvements in certain image areas, but at the same time presents some disadvantages in other retinal zones. As an example, the adoption of the active contour model in the segmentation of wide retinal vessels in [16] involves some disadvantages, as this model can lead to inaccurate segmentations of retinal vessels in some cases of abnormal fundus images.

Since research has shown that the quality of image segmentation depends on the quality of the preprocessing phase, in this paper, the goal of detecting blood vessels in retinal images with DR is reached by developing a Combined Preprocessing Method. The presented technique allows to modify retinal images differently in retinal background and in the optic disc, with the aim of reducing previously mentioned drawbacks towards a screening of proliferative DR. The proposed Combined Preprocessing Method involves more phases as subsequently described in detail. After applying an operator for feature extraction, each retinal vessel image is contrast-enhanced by a contrast-limited adaptive histogram equalization [17, 18]. Then, the suggested Combined Method modifies each contrast-enhanced retinal image both via a two-dimensional matched filtering and via a 2D Gabor wavelet transformation for vessel highlighting. In this way, pairs of modified images containing different vessel informations are obtained, which are properly combined and subsequently segmented. The effectiveness of the proposed method is evaluated on outcomes and results are compared to those obtained with the methods developed in [14] and in [16].

2   Retinal Vessel Detection by a Combined Preprocessing Method

In retinal images the preprocessing phase reveals of a main importance to enhance the contrast between background and retinal vessels, besides removing the effects of improper illumination and noise artifacts. For this purpose, in this section, a Retinal Vessel Detection via a Combined Preprocessing Method is reported as shown in the block diagram reported in Fig. 1, where the input variables of each step are indicated.

After the acquisition of every RGB fundus image, a contrast enhancement step is accurately performed on each gray retinal image I_G. Obtained images are then preprocessed and subsequently properly combined [19]. A Segmentation/Post-processing phase is finally carried out to determine the binary map of blood vessels and reduce noise. In this way each resulting image I_{OUT} is obtained. Each phase is detailed in the following subsections.

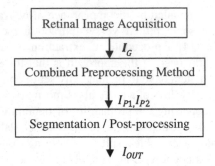

Fig. 1. Block diagram of the proposed Retinal Vessel Detection

2.1 Combined Preprocessing Method

The block diagram of the proposed Combined Preprocessing Method is reported in Fig. 2, where an initial step of Feature Extraction is highlighted. This operation is intended to determine a parameter, called *clip limit β*, which is necessary for the subsequent phase of Contrast Limited Adaptive Histogram Equalization (CLAHE) as in [17, 18].

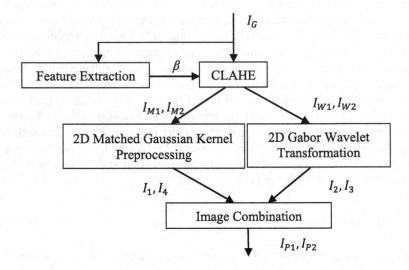

Fig. 2. Block diagram of the combined preprocessing method

Then, an image contrast enhancement step is accurately performed on each gray retinal image I_G to produce two pairs of contrast-enhanced images for every retinal one.

Each couple of obtained contrast-enhanced images I_{M1}, I_{M2} and I_{W1}, I_{W2} is separately preprocessed by a 2D Matched Gaussian Kernel Filtering [10] and a 2D Gabor Wavelet Transformation [16].

Resulting images are finally combined as in [19], determining the preprocessed images I_{P1}, I_{P2} as shown in Fig. 2. The method is subsequently described in detail.

Feature Extraction and Contrast Limited Adaptive Histogram Equalization. The first step of the Combined Preprocessing Method concerns with CLAHE [17, 18, 20, 21]. This technique is a block-based contrast enhancement one, which modifies the dynamic range and the contrast of an image by altering its gray levels, such that its intensity histogram has a desired shape. In detail, it operates on small regions of each retinal image I_G, called *tiles* of size S, to equalize their histograms and control the level of their contrast enhancement by means of the clip limit β, with $0 \leq \beta \leq 1$.

Contrast enhancement is given by the slope of the transformation function in the neighboring of each pixel P of I_G. This slope is, on its turn, proportional to the slope of the Cumulative Distribution Function (CDF) of the neighborhood of P.

The problem of undesirable visual artifacts and possible over-enhancements is overcome by means of the contrast limiting procedure applied to each tile S. Thus, the parameter β limits both the slope of the CDF and the slope of the transformation function obtained for each tile.

Moreover, the choice of parameters of CLAHE is used to achieve variations of the Standard Deviation r and of the Average Value q of the histogram itself, aiming at expanding its distribution on a more large intensity gray level region.

If an image has a very low gray intensity, a high clip limit β makes its histogram flatter and the image becomes brighter, even if noise can be more highlighted.

Furthermore, the contrast of an image can be further modified for a given β using tiles of different size. In the same way, when the size S grows, the dynamic range becomes larger and the contrast of each image increases [20]. Adjacent contrasted tiles of the same size can be finally combined for each image using a bilinear interpolation [21].

Thus, in the proposed Combined Preprocessing Method retinal image preprocessing are improved by selecting tiles of two sizes with the aim of obtaining two values of contrast enhancement for each retinal image.

In this way, four processed images with various values of contrast enhancement are derived using the same clip limit β, each one containing different informations about retinal vessels.

2D Matched Gaussian Kernel Filtering. Blood vessels in human retinal images are generally darker then the background. If $P(x, y)$ indicates any point of the retinal image under exam, the gray intensity profile may be approximated by evaluating the value of a Gaussian curve G in each point P as follows [10]

$$G(P) = a(P)\left\{ 1 - ke^{(-\frac{d^2}{2\sigma^2})} \right\} \tag{1}$$

where the quantity d is the perpendicular distance between the point P and the straight line that passes through the center of each vessel in the direction of its length, the variable σ defines the spread of the gray intensity profile, the quantity $a(P)$ is the gray level intensity of the local background of P and the value k is a measure of reflectance of each blood vessel with reference to its neighborhood.

A 2D-matched Gaussian kernel filtering is herein considered for detecting vessels [10]. This method involves the convolution of a set of Gaussian kernels with the retinal image under investigation, where a generic kernel may be expressed as

$$K(x, y) = -e^{\left(-\frac{x^2}{2\sigma^2}\right)} \quad \text{for } |y| \leq L/2 \tag{2}$$

where L is the length in pixels of the segment for which the vessel is assumed to have a fixed orientation. The negative sign indicates that the vessels are darker than the background. If P' is the generic point in the kernel and φ_i is the orientation of the i-th kernel, with $0 \leq \varphi_i < \pi$ the corresponding point P' (u, v) in the rotated coordinate reference is given by $P'_i = P'R_i^T$ by means of the rotation matrix R_i

$$R_i = \begin{bmatrix} \cos\varphi_i & -\sin\varphi_i \\ \sin\varphi_i & \cos\varphi_i \end{bmatrix} \tag{3}$$

By varying the angle φ of an interval of 15°, a set of 12 kernels is given. A neighborhood N is defined as $N = \{(u, v) | |u| \leq 3\sigma, |v| \leq L/2\}$. The corresponding weights K_i in the i-th kernel are given by

$$K_i(u, v) = -e^{\left(-\frac{u^2}{2\sigma^2}\right)} \quad \forall P'_i \in N \text{ and } |v| \leq L/2 \tag{4}$$

The mean value m_i of each kernel is given by $m_i = \sum_{P'_i \in N} K_i(u, v)/J$ where J denotes the number of points in N. Then the convolutional mask is

$$K'_i(u, v) = K_i(u, v) - m_i \quad \forall P'_i \in N \tag{5}$$

The filter needs to be rotated for all possible angles. Then, the corresponding responses are to be compared and only the maximum one is to be retained for each pixel.

2D-Gabor Wavelet Transformation. The contrast in retinal images is herein enhanced by means of a 2D Gabor wavelet transformation [16]. A continuous wavelet transformation, $T_\psi(\mathbf{b}, \theta, s)$, is determined by the scalar product of a generic image with the transformed wavelet $\psi_{\mathbf{b}, \theta, s}$ as

$$T_\psi(\mathbf{b}, \theta, s) = C_\psi^{-1/2} \langle \psi_{\mathbf{b}, \theta, s} | I_W \rangle = C_\psi^{-\frac{1}{2}} s^{-1} \int \psi^* \left(s^{-1} r_{-\theta}(\mathbf{x} - \mathbf{b})\right) I_W(\mathbf{x}) d^2\mathbf{x} \tag{6}$$

where $\mathbf{x} = (x, y)$, ψ^* is the complex conjugate of ψ, and C_ψ is the normalizing constant. The parameters s, $\mathbf{b}$, and θ denote the dilation scale, displacement vector and rotation angle, respectively. The rotation operator r_θ is given by

$$r_\theta(\mathbf{x}) = (x\cos\theta - y\sin\theta, x\sin\theta + y\cos\theta) \tag{7}$$

where $0 \leq \theta \leq 2\pi$. The 2D Gabor wavelet may be expressed as

$$\psi(\mathbf{x}) = e^{j\mathbf{kx}}e^{(-\frac{1}{2}|B\mathbf{x}|^2)} \tag{8}$$

where $B = diag[\eta^{-1/2}, 1]$ $(\eta \geq 1)$ is a (2×2)-size diagonal matrix, which defines the anisotropy of the Gabor wavelet filter and $\mathbf{k}$ is the complex-exponential frequency vector.

The choice of the quantity η, which controls the elongation of the filter in any desired direction, is critical. In fact, a larger value of η generates more extensive widths of the retinal vessels and a smaller value of η has a less effect on the vessel enhancement. Then, the maximum response $M_\psi(\mathbf{b}, s) = \max\limits_{\theta} |T_\psi(\mathbf{b}, \theta, s)|$ over all possible orientations is extracted for each pixel, where θ is the angle ranging from $0°$ to $170°$, with a step $= 10°$.

Image Combination. In the presented Method, a suitable image combination is performed by a (pixel-by-pixel) multiplication between proper couples of preprocessed images [19] both to enhance the gray levels of the bright pixels (generally vessel ones) and darken corresponding pixels of the background.

In detail, this kind of combination enables to reduce wrong detections of vessel pixels coming from previous steps. In detail, the (pixel-by-pixel) multiplication is performed both between the images I_1 and I_2 and between the images I_3 and I_4. The resulting images I_{P1} and I_{P2}, are obtained as

$$I_{P_1} = I_1(i, j) \times I_2(i, j) \quad I_{P_2} = I_3(i, j) \times I_4(i, j) \tag{9}$$

Images I_{P1} and I_{P2} are then remapped in the range $(0, 255)$ and subsequently segmented.

2.2 Segmentation and Post-processing

The segmentation is performed on each image I_{Pi}, $i = 1, 2$, by applying twice the following method. An initial point $I_{Poi}(x, y)$ is selected for the segmentation of each image I_{Pi}, such that $I_{Poi}(x, y) \geq g_h\text{-}0.05$, where g_h is the gray level intensity corresponding to the 99 % of the histogram size of each image I_{Pi} [16]. Every other pixel $I_{Pi}(x, y)$ can be related to $I_{Poi}(x, y)$ by the relation

$$|I_{Poi}(x, y) - I_{Pi}(x, y)| < e \tag{10}$$

where the value e derives from the gray level distribution of each image I_{Pi}.

If (10) is satisfied, the considered pixel $I_{Pi}(x, y)$ belongs to a vessel, otherwise it belongs to background. The segmentation is carried out for all pixels of I_{P1}, I_{P2}. Obtained segmented images are subsequently combined by means a logical OR operation. Proper steps are finally performed to remove falsely detected isolated vessels pixels, providing image I_{OUT}.

3   Experimental Results and Discussion

The proposed Combined Preprocessing Method has been applied to the seven retinal images with DR taken from the publicly available DRIVE database [7], together with the corresponding Ground Truth images.

The database DRIVE consists of 40 fundus images taken from a Dutch screening program for diabetic retinopathy, where 33 images are of healthy retinas and the other 7 show signs of mild early proliferative diabetic retinopathy [14]. All these images were captured in digital form from a Canon CR5 nonmydriatic 3CCD camera at 45 field of view (FOV). They are (565 × 584)-pixel sized, 8 bit per color channel and have a FOV of approximately 540 pixels in diameter.

A mask image is provided for each image to delineate the FOV. All images were manually segmented by three eye doctors. The 40 images are separated into a training set and a test one, each containing 20 images. Retinal images reported in the test set in DRIVE have been segmented twice, resulting in a Ground Truth set A and a Ground Truth set B. In this paper, the segmentation of set A has been considered as the reference Ground Truth set. For each image I_G under investigation, the feature extraction process has provided values of the clip limit β belonging to the range [0.03–0.05], depending on both the values of the standard deviation r and of the average value q obtained from the histogram in the FOV.

In detail, by considering r^* as the average value of the standard deviations r and q^* as the average value among the values q for all considered images, respectively, it follows that

• If $q > q^*$,	then $\beta = \beta'$
• If $q \leq q^*$ and $r \leq r^*$,	then $\beta = \beta^*$
• If $q \leq q^*$ and $r > r^*$,	then $\beta = \beta''$

where the clip limit β^* is the average value in the considered range, β'' and β' are obtained by varying β^* of $\pm\, 25\,\%$, respectively.

Moreover, tiles of different sizes have been selected for contrast enhancing pathologic retinal images to divide each original image I_G in (8 × 8) tiles and (50 × 50) tiles, respectively. Then, the quantities L and σ have been adopted as in [10] for the 2D Matched Gaussian Kernel Filtering; the variables η, $\mathbf{k}$, s have been considered as in [16], when performing the 2D-Gabor Wavelet Transformation to obtain a reliable comparison with the corresponding methods.

The presented method is herein evaluated quantitatively on all DR images selected from the DRIVE database. Each image of the set is preprocessed according to the above procedure to obtain the corresponding I_{OUT}.

For the sake of a better comprehension, the Combined Preprocessing Method is applied to a selected retinal image I_{G19} and the corresponding images obtained in intermediate phases are reported in Fig. 3.

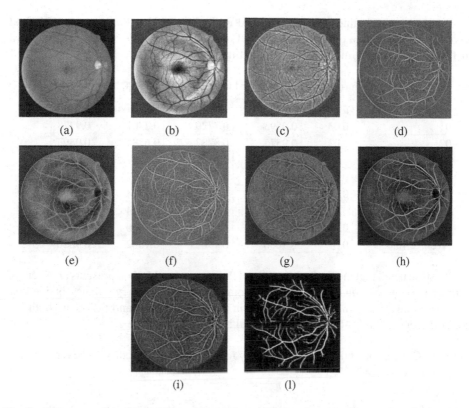

Fig. 3. Outcomes of the Retinal Vessel Detection applied to the retinal image I_{G19}: (a) I_{G19}; (b) I_{M1}; (c) I_{M2}; (d) I_1; (e) I_2; (f) I_4; (g) I_3; (h) I_{P1}; (i) I_{P2}; (l) I_{OUT}

The performances of the presented method have been then investigated by considering the variable Accuracy, which is strongly related to the segmentation quality and it is often used to evaluate and compare different methods.

In detail, the quantity Accuracy is defined as the ratio of correctly classified vessel pixels and non-vessel ones to the total number of pixels in FOV. This quantity represents the most commonly adopted index for performance evaluation, since it is strongly related to the segmentation quality of images.

Moreover, other statistical quantities, such as Precision, Sensitivity and Specificity, which indicate features regarding with binary segmented outcomes, have been computed to characterize the proposed method.

In detail, Precision is given by the ratio of the pixels correctly classified as vessel pixels to the total number of pixels classified as vessel ones; Sensitivity is defined as the number of pixels correctly classified as vessel pixels divided by the number of vessel ones in the corresponding Ground Truth image, whereas Specificity is determined by the ratio of the number of pixels correctly classified as background ones to the number of background pixels in the corresponding Ground Truth one.

Performances obtained for the proposed Retinal Combined Preprocessing Method have been expressed in terms of Accuracy, Precision, Sensitivity, Specificity as shown in Table 1, together with the values of clip limit β.

Table 1. Performance of the proposed method on retinal images with PDR

Image	β	Accuracy	Precision	Sensitivity	Specificity
01	0.03	0.9593	0.7729	0.7701	0.9778
02	0.03	0.9575	0.8787	0.6784	0.9893
10	0.04	0.9597	0.8352	0.6357	0.9888
14	0.03	0.9563	0.7264	0.7367	0.9756
18	0.03	0.9610	0.7548	0.7514	0.9790
19	0.05	0.9688	0.8378	0.7744	0.9864
20	0.04	0.9615	0.7258	0.7650	0.9771
Average value	0.04	0.9606	0.7902	0.7302	0.9820

It can be noted that the Retinal Combined Preprocessing Method presents an Average Value of Accuracy equal to 96,06 %. A comparison of Accuracy values obtained by preprocessing retinal images with the herein proposed method and following both the method proposed in [14] and the one in [16] is reported in Table 2.

Table 2. Comparison of the values of accuracy for retinal vessel detection on the same images with different methods

Method	Image						
	1	2	10	14	18	19	20
[14]	0.9384	0.9373	0.9372	0.9406	0.9373	0.9489	0.9400
[16]	0.9500	0.9498	0.9457	0.9517	0.9501	0.9618	0.9550
Proposed	0.9593	0.9575	0.9597	0.9563	0.9610	0.9688	0.9615

It can be noted that better values of Accuracy are obtained with the proposed method, when compared with the methods developed in [14] and in [16], and lower values of clip limit β for all the retinal images under investigation have been used.

4   Conclusions

In this paper a new Combined preprocessing method for retinal vessel detection towards a proliferative DR screening has been presented and developed. The effectiveness of the proposed method has been evaluated by considering diseased retinal images available in a publicly database for scientific analysis of DR. Obtained outcomes and results have been compared to those obtained with two significant methods. Experimental results show that better values of Accuracy have been obtained with the presented method with respect to the others. Future work could be devoted to perform a dynamic selection of the values of clip limit and size of tiles for preprocessing each damaged retinal image. Finally, the innovative method herein proposed for retinal vessel detection towards a proliferative DR screening could enable the identification of previously unrecognized novel markers of disease threat.

Acknowledgement. The authors' acknowledgement goes to the financial support to this research given by the F.R.A. 2012 Fund - Technical University of Bari.

References

1. Singh, N., Kaur, L.: A survey on blood vessel segmentation methods in retinal images. In: Proceedings of International Conference on Electronic Design, Computer Networks and Automated Verification (EDCAV), pp. 23–28 (2015)
2. Dash, J., Bhoi, N.: A survey on blood vessel detection methodologies in retinal images. In: Proceedings of 2015 International Conference on Computational Intelligence and Networks (CINE), pp. 166–171 (2015)
3. Bevilacqua, V., Pietroleonardo, N., Triggiani, V., Gesualdo, L., Di Palma, A.M., Rossini, M., Dalfino, G., Mastrofilippo, N.: Neural network classification of blood vessels and tubules based on Haralick features evaluated in histological images of kidney Biopsy. In: Huang, D.-S., Han, K. (eds.) ICIC 2015. LNCS, vol. 9227, pp. 759–765. Springer, Heidelberg (2015)
4. Bevilacqua, V.: Three-dimensional virtual colonoscopy for automatic polyps detection by artificial neural network approach: new tests on an enlarged cohort of polyps. Neurocomputing **116**, 62–75 (2013)
5. Abramoff, M.D., Garvin, M., Sonka, M.: Retinal image analysis: a review. IEEE Rev. Biomed. Eng. **3**, 169–208 (2010)
6. Sim, D.A., Keane, P.A., Tufail, A., Egan, C.A., Aiello, L.P., Silva, P.S.: Automated retinal image analysis for diabetic retinopathy in telemedicine. Microvascular Complications—Retinopathy (J.K. Sun, Section ed.) Current Diabetes Reports **15**(3) (2015). Springer US
7. Staal, J., Abràmoff, M.D., Niemeijer, M., Viergever, M.A., van Ginneken, B.: Ridge-based vessel segmentation in color images of the retina. IEEE Trans. Med. Imaging **23**(4), 501–509 (2004)
8. Carnimeo, L., Bevilacqua, V., Cariello, L., Mastronardi, G.: Retinal vessel extraction by a combined neural network–wavelet enhancement method. In: Huang, D.-S., Jo, K.-H., Lee, H.-H., Kang, H.-J., Bevilacqua, V. (eds.) ICIC 2009. LNCS, vol. 5755, pp. 1106–1116. Springer, Heidelberg (2009)
9. Carnimeo, L., Benedetto, A.C., Mastronardi, G.: A voting procedure supported by a neural validity classifier for optic disk detection. In: Huang, D.-S., Gupta, P., Zhang, X., Premaratne, P. (eds.) ICIC 2012. CCIS, vol. 304, pp. 467–474. Springer, Heidelberg (2012)
10. Chaudhuri, S., Chatterjee, S., Katz, N., Nelson, M., Goldbaum, M.: Detection of blood vessels in retinal images using two-dimensional matched filters. IEEE Trans. Med. Imaging **8**(3), 263–269 (1989)
11. Hoover, A., Kouznetsova, V., Goldbaum, M.: Locating blood vessels in retinal images by piecewise threshold probing of a matched filter response. IEEE Trans. Med. Imaging **19**(3), 203–210 (2000)
12. Bevilacqua, V., Carnimeo, L., Mastronardi, G., Santarcangelo, V., Scaramuzzi, R.: On the comparison of NN-based architectures for diabetic damage detection in retinal images. J. Circ. Syst. Comput. **18**(8), 1369–1380 (2009)
13. Zhang, D., Li, Q., You, J., Zhang, D.: A modified matched filter with double-sided thresholding for screening proliferative diabetic retinopathy. IEEE Trans. Inf. Technol. Biomed. **13**(4), 528–534 (2009)

14. Ramlugun, G.S., Nagarajan, V.K., Chakraborty, C.: Small retinal vessels extraction towards proliferative diabetic retinopathy screening. Expert Syst. Appl. **39**(1), 1141–1146 (2012). (Elsevier)
15. Carnimeo, L., Nitti, R.: On classifying diabetic patients' with proliferative retinopathies via a radial basis probabilistic neural network. In: Huang, D.-S., Han, K. (eds.) ICIC 2015. LNCS, vol. 9227, pp. 115–126. Springer, Heidelberg (2015)
16. Zhao, Y.Q., Wang, X.H., Wang, X.F., Shih, F.Y.: Retinal vessels segmentation based on level set and region growing. Pattern Recogn. **47**(7), 2437–2446 (2014). (Elsevier)
17. Zuiderveld, K.: Contrast limited adaptive histogram equalization (VIII. 5). In: Heckbert, P.S. (ed.) Graphics Gems IV, pp. 474–485. Academic Press, Cambridge (1994)
18. Reza, A.M.: Realization of the Contrast Limited Adaptive Histogram Equalization (CLAHE) for real-time image enhancement. J. VLSI Signal Process. Syst. Signal Image Video Technol. **38**(1), 35–44 (2004)
19. Umbaugh, S.E.: Digital Image Processing and Analysis: Human and Computer Vision Applications with CVIPtools, 2nd edn. CRC Press, Boca Raton (2010)
20. Min, B.S., Lim, D.K., Kim, S.J., Lee, J.H.: A novel method of determining parameters of CLAHE based on image entropy. Int. J. Softw. Eng. Appl. **7**(5), 113–120 (2013)
21. Vyas, G., Thakur, A., Bhan, A.: Analysis of histogram based contrast enhancement with noise reduction method for endodontic therapy. In: Proceedings of the 3rd International Conference on Reliability, Infocom Technologies and Optimization (ICRITO) (Trends and Future Directions), pp. 1–5 (2014)

A New Flexible Protocol for Docking Studies

Lucia Sessa[1](✉), Luigi Di Biasi[1,2], Simona Concilio[3], Giuseppe Cattaneo[2],
Alfredo De Santis[2], Pio Iannelli[1], and Stefano Piotto[1]

[1] Department of Pharmacy, University of Salerno, Via Giovanni Paolo II, 132,
84084 Fisciano, SA, Italy
lucsessa@unisa.it
[2] Department of Informatics, University of Salerno, Via Giovanni Paolo II, 132,
84084 Fisciano, SA, Italy
[3] Department of Industrial Engineering, University of Salerno, Via Giovanni Paolo II, 132,
84084 Fisciano, SA, Italy

Abstract. A significant prerequisite for computational structure-based drug
design is the estimation of the structures of ligand-receptor complexes. For this
task, the flexibility of both ligand and receptor backbone is required, but it requires
the exploration of an extremely vast conformational space. Here we present a
protocol to address the receptor flexibility using complementary strategies and
the use of receptor sequence conservation. The method aims to increase the accuracy of predicted ligand orientation in the binding pocket and the receptor-ligand
binding affinity. The precision in affinity prediction permits to distinguish
between binders and non-binders and to identify binding sites and ligand poses
necessary for lead optimization.

1 Introduction

It is very important to estimate the in silico potential toxicity of existing or hypothetical
compounds. This can be achieved simulating the interactions of a ligand towards target
proteins suspected to trigger adverse effect. Docking methods can be useful to evaluate
potential toxicity of small molecule [1], and to distinguish between binders and non-
binders. Calculation of free energy of binding is a good way to estimate the binding
affinity between two molecules. We used the enzyme Androgen Receptor (AR), a
nuclear receptor activated by binding androgenic hormones, testosterone, or dihydro-
testosterone to test the flexible protocol. AR plays an essential role in the growth of
normal prostate, and it is involved in the development of prostate cancer [2]. When the
experimental structure of the complex ligand-receptor is known, the ligand can be
docked directly in the binding site. However, experimental and theoretical studies show
that proteins can fluctuate between different conformations in the absence of ligand
[3–9]. In aqueous solution, proteins domains are in constant motion exhibiting a conformational heterogeneity [10–15]. Molecular recognition involves non-covalent association of ligands to protein target with high affinity and specificity. According to the
Koshland's 'induced-fit' model, the interaction between a protein and a ligand induces
a conformational change in the protein [16]. Among the conformations of the dynamically fluctuating protein, the receptor selects the one that better accommodates the

© Springer International Publishing Switzerland 2016
F. Rossi et al. (Eds.): WIVACE 2015, CCIS 587, pp. 117–126, 2016.
DOI: 10.1007/978-3-319-32695-5_11

ligand. In case of modeling protein-ligand complexes, it is necessary to consider back-bone movement. The major flaw in docking methods is to consider only a single repre-sentative structure for the receptor. Introducing receptor flexibility in a standard docking protocol is a way to study conformational changes. Flexibility is particularly important when the binding pocket is inside the receptor and the ligand interactions induce struc-tural movements of the backbone of the receptor. We explored the conformation vari-ability by addressing the receptor flexibility and taking into account the receptor sequence conservation.

Keeping proteins flexible during the docking has a high computational cost in virtue of the high number of degrees of freedom. To overcome this limit, our docking strategy is to represent receptor flexibility by utilizing an inexpensive method that offers more target structures.

First, we considered several side-chain conformations of the receptors and we calcu-lated, for each rotamer, the values of binding energy of a ligand during a molecular dynamics simulation. In a second approach, we create a docking ensemble of structures derived by steered molecular dynamics (SMD). In this method, we collected the trajec-tory when the ligand is pulled away from its binding pocket. SMD simulation is an inexpensive computational method and offers more target structures to perform docking. In the third approach, we consider the low frequency vibrations of a protein. Protein low-frequency vibrations retain important biological functions [17]. For each vibrational mode, we generated a set of snapshots in a similar fashion to the first approach. Finally, we present a protocol for docking of ligands into flexible protein binding sites. The protocol was tested on a list of therapeutically relevant targets with available crystallo-graphic data. The free energy was calculated as the difference between the energy of separated compounds and the energy of the complex. However, the computational complexity of the procedure grows quickly with the numbers of conformers considered. Consequently, to reduce the computational time and cost we have used a specialized grid (GRIMD) to distribute all jobs [18].

2   Materials and Methods

2.1   Data Set – Selection of Complexes

The structures for the docking simulations were taken from the X-ray structures in the PDB database [19]. We performed a search using the following query parameters: human protein Androgen receptor (Uniprot ID P10275), structure complexed with a ligand containing experimental binding affinity and the X-ray resolution up to 2.0 Å to ensure crystallographic structures with sufficient structural quality.

The receptor structures were prepared by removing all water molecules and substrates including the ligand molecule. All structures were prepared adding missing hydrogens and optimizing the hydrogen-bonding network. The internal cavity volume inside the macro-molecules was calculated with the software YASARA Structure 15.6.21 [20].

2.2   Classical MD Simulations

We represented receptor flexibility by utilizing several snapshot from molecular dynamics (MD). The MD simulations were performed with the software YASARA Structure 15.6.21. We used AMBER03 as force field with long-ranged PME potential and a cutoff of 8.0 Å. The starting structures for the simulations of the ARs were extracted from the X-ray structures from the PDB database. A cubic periodic simulation cell of 512000 Å^3 was defined around all atoms of the receptor structures. The MD simulation was then initiated at 298 K and integration time steps for intramolecular forces every 1.25 fs.

Ten structures for each receptor were selected from the MD simulation at regular time intervals (each 500 ps). A simulation cell was centered on the binding pocket sides and dimensions of the box were adapted for each structure to cover the entirety of the active site.

2.3   Steered Molecular Dynamics Simulation

We represented receptor flexibility utilizing structures collected by steered molecular dynamics simulation (SMD). The SMD was carried out using the software YASARA Structure 15.6.21. We collected the trajectory traced when the ligand is pulled from its binding pocket to the outside. The initial structures were retrieved from the X-ray structures from the PDB database and were cleaned by erasing all water molecules and substrates different from the ligand molecule. A cubic periodic simulation cell of 512000 Å^3 was built around the entire complex. The charges were assigned at physiological conditions (pH 7.4). The simulation box was filled with water choosing a density of 0.997 g/mL. The simulation cell was neutralized with NaCl with a final concentration of 0.9 %. We minimized the energy of the system using first a steepest descent minimization followed by a simulated annealing minimization. The pulling acceleration of the ligand was 3 Å/ps^2. The simulation was stopped when the distance between the centers of mass of receptor and ligand was > 30 Å.

2.4   Low Vibrational Modes

We considered the low frequency vibrations that a protein may undergo. The lowest vibrational modes of a protein were determined using the Schrodinger's biologics suite [15]. Before running the calculation, the proteins were properly prepared, using the Protein Preparation Wizard and all waters and solvent molecules were deleted. The input structures were minimized and the vibrational modes were generated as a set of structures sampled at regular intervals along a full cycle of the vibrational mode. We set the following parameters to 5 vibrational modes to view, and 20 frames per mode.

For each vibrational mode, we generated a set of snapshots of the structure at a particular point in the vibration. On these structures, we performed docking simulation similarly to the MD approach.

2.5   Docking Method

Ligand structures were extracted from the X-ray pose of the complex receptor-ligands and were minimized into YASARA by using AMBER03 force field.

The molecular docking simulations were performed using VINA provided in the YASARA package. The force field selected was AMBER03. The ligands were independently docked 250 times against 5 receptor ensembles with alternative high-scoring solutions of the side-chain rotamer network each. The results were clustered using a tolerance of 5.0 Å. We calculated the values of binding energy and corresponding binding constants of each ligand respect to the receptor.

In addition, we used a new tool for molecular docking, named mod-VINA, developed from a modification of AutoDock VINA [21]. This method permits to increase the accuracy of docking and it is useful to predict the best ligand pose correlating the docking analysis to the receptor sequence conservation. Contrarily to VINA, this tool does not create the ligand poses in a random manner in all the space available around the receptor, but it generates a pose in local spheres created close to conserved residues. With this approach, mod-VINA not only improves its accuracy in pose generation, but also reduces considerably the computing time elapsed for blind docking. mod-VINA is convenient when the binding pocket is unknown and the only strategy is to perform a blind docking.

3   Results and Discussion

We collected several snapshots of receptor structure by MD, SMD and vibrational low modes.

To validate the accuracy of our flexible protocol in docking, we have used a set of 10 crystallographic conformations of AR with the same target but with a different ligand. In Table 1 are reported the PDB codes of the selected receptors and the corresponding ligands.

Table 1.   The AR targets and the corresponding ligand

PDB code for AR	Ligand
2AM9	TES
2AX6	HFT
2AX8	FHM
2AX9	BHM
3B5R	B5R
3B65	3B6
3B66	B66
3B67	B67
3B68	B68
3L3X	DHT

The collected proteins showed the 100 % of sequence identity and the same position of the buried binding pocket.

We performed two parallel experiments of docking, one with a rigid protein target and one considering flexible receptor structures. In addition, we compared the results for both experiments in the re-docking and in the cross-docking analysis.

3.1 Rigid Receptor Docking Simulations

The predicted structure of a ligand-receptor complex with the better binding energy was superposed on the crystallographic complex conformation to calculate the root mean square deviation (RMSD).

The RMSD values are reported in Fig. 1. The cell holding the minimum value of RMSD (0.5 Å) is colored in green and the cell with the maximum value of RMSD (20 Å) is colored in white. All other cells are colored proportionally.

Diagonals values are the RMSD values for the re-docking analysis. As expected, the values are very low and accurate. These results were not surprising because the receptor structure used was exactly the conformation adopted in the bound state from the receptor with the specific ligand.

In the cross-docking studies, we docked the ligand from one complex into the receptor of the other complex and *vice versa*. The RMSD values between the predicted poses and the experimental data are reported in Fig. 1 and are color mapped as described before for the re-docking.

	2AM9	2AX6	2AX8	2AX9	3B5R	3B65	3B66	3B67	3B68	3L3X
TES	0.6	6.7	14.3	4.4	14.6	17.0	6.7	6.7	6.7	0.7
HFT	1.9	1.3	15.1	1.1	1.1	1.2	1.2	1.0	1.1	14.6
FHM	15.2	15.9	0.9	16.1	1.1	1.7	1.7	0.9	0.8	16.5
BHM	2.0	1.4	15.1	1.0	1.2	1.1	1.2	1.4	1.4	2.1
B5R	9.1	16.1	1.3	16.4	1.0	1.0	1.8	1.8	1.8	17.0
3B6	9.5	15.4	1.8	17.3	1.7	1.6	1.7	1.8	1.7	17.6
B66	9.4	17.1	1.9	17.1	1.0	0.8	1.8	1.8	1.7	17.4
B67	15.1	15.9	1.0	16.1	1.2	2.0	1.9	0.7	0.9	16.2
B68	18.6	13.9	1.6	17.0	1.9	1.7	1.8	1.9	1.0	18.8
DHT	0.6	6.7	14.2	4.2	14.6	15.2	6.6	6.6	6.6	0.7

RMSD (Å) scale: 0   4   8   12   16   20

Fig. 1. Re-docking and cross-docking results for 10 different conformations of AR. The color mapped RMSD values are calculated between the predicted pose with highest energy of binding and the experimental pose (Color figure online).

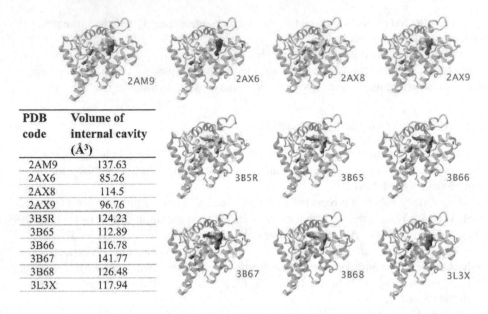

PDB code	Volume of internal cavity (Å³)
2AM9	137.63
2AX6	85.26
2AX8	114.5
2AX9	96.76
3B5R	124.23
3B65	112.89
3B66	116.78
3B67	141.77
3B68	126.48
3L3X	117.94

Fig. 2. Internal cavities of 10 different X-ray poses for AR. The secondary structures of the receptors are colored in gray to put in evidence the shape of their internal cavities (Color figure online).

We found that rigid cross-docking fails with several types of complexes with values of RMSD up to 18 Å. Analyzing the docked poses we observed that in those cases, the most favorite position for the ligand is on the receptor molecular surface and not in the empty space inside the receptor.

To understand this docking mistake, we measured the binding pocket buried inside the receptor calculating the volume of the internal cavities.

Figure 2 shows the internal cavity of the ten receptors and the volume for each cavity. It is possible to observe that AR displays different internal cavities shapes and volumes.

Slight changes in the structure of the binding pocket between different crystallographic structures can radically affect the outcome and alter the docking results. As showed for the receptors 3L3X or 2AX9 the internal cavity of the buried binding pocket does not have space enough to accommodate all ligands. In these cases, the cross-docking results are very bad. The changes of the cavity volume might explain the failure of traditional docking method and support the hypothesis that a single representative structure for the receptor is not enough.

3.2   Flexibility Receptor Docking Simulations

The comparison among the different approaches is summarized in Fig. 3.

We show the results for two different X-ray poses of AR (3B68 and 3L3X) re-docked with their crystallographic ligand such as B68 and DHT respectively (A and D graphs) and with a second substrate (B and C graphs). We reported the RMSD calculated

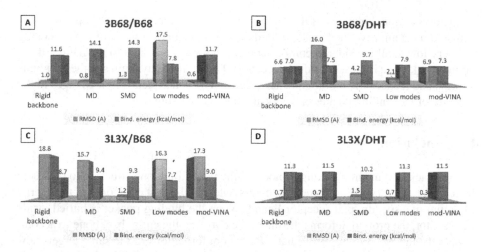

Fig. 3. Flexible protocol results for AR structures 3B68 and 3L3X re-docked (A and D) and cross-docked (B and C) with B68 and DHT. Here we show the docked poses with the highest energy of binding (blue column) predicted using the rigid backbone taken from the X-ray structure and the snapshots from MD, SMD and low-frequency vibrations. In addition we report the results obtained using the tool mod-VINA. To establish the accuracy of the prediction we reported the RMSD values (green column) calculated between the docked pose and the experimental one (Color figure online).

between the experimental and the predicted pose with the highest energy of binding. Low RMSD values indicate the correct superpose between two structures, also the more positive values of binding energy are correlated to the more favorable receptor-ligand interaction.

The re-docking experiments (A and D graphs) are very accurate with RMSD < 1 Å. As expected, in these cases the flexible docking protocol does not improve the results. This was not true for the cross-docking simulations (B and C graphs) where a flexible docking protocol can improve the overall quality of the simulations. In the receptor conformation in 3L3X, the internal cavity does not have space enough to accommodate the ligand B68, so the docking results are poor. However, we improved the docking performances using the SMD approach. We observed an RMSD value of 1.2 Å that means a complete overlapping between the cross-docked pose of B68 docked in the AR conformation taken from the 3L3X PDB and the experimental complex structure reported in the 3B68 PDB.

The other system analyzed was formed by the receptor structure frozen in 3B68 and the ligand DHT. This receptor shows a volume of the internal cavity big enough to host the substrate DHT. However, the overall quality (Fig. 3) is not satisfactory. In fact, the shape of the cavity may not be compatible with the geometry of the ligand. We obtained good results using the snapshot collected by the low-frequency vibrations in macromolecules. The different conformations collected from the low motions included a compatible internal cavity in term of shape and volume resulting in a good cross-docking performance (RMSD 2.1 Å).

In contrast, the dynamics of the receptor backbone, an approach conceptually very simple, did not improve the results of rigid docking. The performance of cross-docking was very low in all considered conformations. This was attributed to the nature of the binding pocket that is buried inside the receptor. Therefore the dynamics of the backbone deforms and reduces the binding pocket volume. The consequent slight changes in the structure of the binding pocket radically affected the docking results.

4   Conclusion

A major lack in standard docking protocol is the use of only one structure to represent the receptor. The crystallographic data corresponds, in the best case, only to the energy minimum of a specific pair receptor-ligand. When the binding pocket is buried inside the receptor, the protein deforms the internal cavity and the entire backbone to accommodate the ligands. We observed that different conformations of the same target show different volume and shape of the internal cavity. Consequently, the cross-docking analysis are typically very poor. To increase the accuracy of cross-docking we suggest a flexible protocol to deform the internal cavity in a "natural" way. We increased the receptor flexibility by enhancing target information and improving the precision of ligand position. We represented receptor flexibility by utilizing several snapshot from molecular dynamics simulations (MD), by steered molecular dynamics simulation (SMD) and using the structures extracted from the low frequency vibrations that a protein may undergo. Supposing that the binding pocket in a receptor corresponds to a conserved receptor region, we used a new tool (mod-VINA) for the docking developed from VINA to correlate the cross-docking simulation to the receptor sequence conservation.

The flexible model was used for cross-docking tests of 10 ligands and 10 different conformations of the androgen receptor. Based on the docking results we suggest that the SMD method provides various conformations with different internal cavities that can host ligands of different geometry. The low-vibrational mode dynamics was another good way to collect receptor structures. The protein vibrations were able to deform the internal cavity in a non-invasive way, as demonstrated by the good results in cross-docking experiment. In contrast, the molecular dynamics simulations were too strong in deforming the cavity of the binding pocket. As result, the cavity collapses and the docked ligands were placed outside the binding pocket on the receptor surface. In the fourth approach we taken into account the protein conservation. In the androgen receptor, the binding pocket is completely conserved. The best poses predicted by mod-VINA were completely superposed to the experimental structure for the re-docking results. Anyway, in the cross-docking this approach was not efficient, because of the different size and shape of the binding pockets.

The flexible docking protocol described in this work is a simple way to generate multiple receptor conformations that can be used to easily dock a large number of compounds.

Acknowledgments. This work was partially supported by the "Data-Driven Genomic Computing (GenData 2020)" PRIN project (2013–2015), funded by the Italian Ministry of the University and Research (MIUR).

References

1. Goldsmith, M.-R., Grulke, C.M., Chang, D.T., Transue, T.R., Little, S.B., Rabinowitz, J.R., Tornero-Velez, R.: DockScreen: A Database of In Silico Biomolecular Interactions to Support Computational Toxicology. Dataset Papers in Science **2014** (2014)
2. Pomerantz, M.M., Li, F., Takeda, D.Y., Lenci, R., Chonkar, A., Chabot, M., Cejas, P., Vazquez, F., Cook, J., Shivdasani, R.A., Bowden, M., Lis, R., Hahn, W.C., Kantoff, P.W., Brown, M., Loda, M., Long, H.W., Freedman, M.L.: The androgen receptor cistrome is extensively reprogrammed in human prostate tumorigenesis. Nat. Genet. **47**(11), 1346–1351 (2015)
3. Lopez, D.H., Fiol-deRoque, M.A., Noguera-Salvà, M.A., Terés, S., Campana, F., Piotto, S., Castro, J.A., Mohaibes, R.J., Escribá, P.V., Busquets, X.: 2-Hydroxy arachidonic acid: a new non-steroidal anti-inflammatory drug. PLoS ONE **8**(8), e72052 (2013)
4. Piotto, S., Concilio, S., Bianchino, E., Iannelli, P., López, D.J., Terés, S., Ibarguren, M., Barceló-Coblijn, G., Martin, M.L., Guardiola-Serrano, F.: Differential effect of 2-hydroxyoleic acid enantiomers on protein (sphingomyelin synthase) and lipid (membrane) targets. Biochim. Biophys. Acta (BBA)-Biomembr. **1838**(6), 1628–1637 (2014)
5. Piotto, S., Concilio, S., Mavelli, F., Iannelli, P.: Computer simulations of natural and synthetic polymers in confined systems. Macromolecular symposia **286**(1), 25–33 (2009)
6. Piotto, S., Trapani, A., Bianchino, E., Ibarguren, M., López, D.J., Busquets, X., Concilio, S.: The effect of hydroxylated fatty acid-containing phospholipids in the remodeling of lipid membranes. Biochim. Biophys. Acta (BBA)- Biomembr. **1838**(6), 1509–1517 (2014)
7. Piotto, S.P., Sessa, L., Concilio, S., Iannelli, P.: YADAMP: yet another database of antimicrobial peptides. Int. J. Antimicrob. Agents **39**(4), 346–351 (2012)
8. Scrima, M., Di Marino, S., Grimaldi, M., Campana, F., Vitiello, G., Piotto, S.P., D'Errico, G., D'Ursi, A.M.: Structural features of the C8 antiviral peptide in a membrane-mimicking environment. Biochim. Biophys. Acta (BBA)- Biomembr. **1838**(3), 1010–1018 (2014)
9. Piotto, S., Nesper, R.: CURVIS: A program to study and analyse crystallographic structures and phase transitions. J. Appl. Crystallogr. **38**(1), 223–227 (2005)
10. Torok, Z., Crul, T., Maresca, B., Schutz, G.J., Viana, F., Dindia, L., Piotto, S., Brameshuber, M., Balogh, G., Peter, M., Porta, A., Trapani, A., Gombos, I., Glatz, A., Gungor, B., Peksel, B., Vigh Jr., L., Csoboz, B., Horvath, I., Vijayan, M.M., Hooper, P.L., Harwood, J.L., Vigh, L.: Plasma membranes as heat stress sensors: from lipid-controlled molecular switches to therapeutic applications. Biochim. Biophys. Acta **1838**(6), 1594–1618 (2014)
11. Crul, T., Toth, N., Piotto, S., Literati-Nagy, P., Tory, K., Haldimann, P., Kalmar, B., Greensmith, L., Torok, Z., Balogh, G., Gombos, I., Campana, F., Concilio, S., Gallyas, F., Nagy, G., Berente, Z., Gungor, B., Peter, M., Glatz, A., Hunya, A., Literati-Nagy, Z., Vigh Jr., L., Hoogstra-Berends, F., Heeres, A., Kuipers, I., Loen, L., Seerden, J.P., Zhang, D., Meijering, R.A., Henning, R.H., Brundel, B.J., Kampinga, H.H., Koranyi, L., Szilvassy, Z., Mandl, J., Sumegi, B., Febbraio, M.A., Horvath, I., Hooper, P.L., Vigh, L.: Hydroximic acid derivatives: pleiotropic HSP co-inducers restoring homeostasis and robustness. Curr. Pharm. Des. **19**(3), 309–346 (2013)

12. Gombos, I., Crul, T., Piotto, S., Gungor, B., Torok, Z., Balogh, G., Peter, M., Slotte, J.P., Campana, F., Pilbat, A.M., Hunya, A., Toth, N., Literati-Nagy, Z., Vigh Jr., L., Glatz, A., Brameshuber, M., Schutz, G.J., Hevener, A., Febbraio, M.A., Horvath, I., Vigh, L.: Membrane-lipid therapy in operation: the HSP co-inducer BGP-15 activates stress signal transduction pathways by remodeling plasma membrane rafts. PLoS ONE **6**(12), e28818 (2011)

13. Vigh, L., Torok, Z., Balogh, G., Glatz, A., Piotto, S., Horvath, I.: Membrane-regulated stress response: a theoretical and practical approach. In: Vigh, L., Csermely, L. (eds.) Molecular Aspects of the Stress Response: Chaperones, Membranes and Networks. Advances in experimental medicine and biology, vol. 594, pp. 114–131. Springer Science + Business Media, New York (2007)

14. Lange, O.F., Lakomek, N.-A., Farès, C., Schröder, G.F., Glatz, A., Walter, K.F., Becker, S., Meiler, J., Grubmüller, H., Griesinger, C., De Groot, B.L.: Recognition dynamics up to microseconds revealed from an RDC-derived ubiquitin ensemble in solution. Science **320**(5882), 1471–1475 (2008)

15. Mittermaier, A., Kay, L.E.: New tools provide new insights in NMR studies of protein dynamics. Science **312**(5771), 224–228 (2006)

16. Tuffery, P., Derreumaux, P.: Flexibility and binding affinity in protein–ligand, protein–protein and multi-component protein interactions: limitations of current computational approaches. J. Roy. Soc. Interface **9**(66), 20–33 (2012)

17. Chou, K.C.: Biological functions of low-frequency vibrations (phonons). III. Helical structures and microenvironment. Biophys. J. **45**(5), 881–889 (1984)

18. Piotto, S., Biasi, L.D., Concilio, S., Castiglione, A., Cattaneo, G.: GRIMD: distributed computing for chemists and biologists. Bioinformation **10**(1), 43–47 (2014)

19. Berman, H.M., Westbrook, J., Feng, Z., Gilliland, G., Bhat, T., Weissig, H., Shindyalov, I.N., Bourne, P.E.: The protein data bank. Nucleic Acids Res. **28**(1), 235–242 (2000)

20. Krieger, E., Vriend, G.: YASARA View - molecular graphics for all devices - from smartphones to workstations. Bioinformatics **30**(20), 2981–2982 (2014)

21. Trott, O., Olson, A.J.: AutoDock Vina: improving the speed and accuracy of docking with a new scoring function, efficient optimization, and multithreading. J. Comput. Chem. **31**(2), 455–461 (2010)

Design and Development of a Forearm Rehabilitation System Based on an Augmented Reality Serious Game

Vitoantonio Bevilacqua[1]([⊠]), Antonio Brunetti[1], Giuseppe Trigiante[1],
Gianpaolo Francesco Trotta[1], Michele Fiorentino[2], Vito Manghisi[2],
and Antonio Emmanuele Uva[2]

[1] Dipartimento di Ingegneria Elettrica e dell'Informazione, Politecnico di Bari,
via Orabona 4, 70125 Bari, Italy
vitoantonio.bevilacqua@poliba.it
[2] Dipartimento di Ingegneria Meccanica, Matematica e Management,
Politecnico di Bari, viale Japigia 182, 70126 Bari, Italy

Abstract. In this paper, we propose a forearm rehabilitation system based on a serious game in Augmented Reality (AR). We designed and developed a simplified AR arcade brick breaking game to induce rehabilitation of the forearm muscles. We record the electromyographic signals using a low cost device to evaluate the applied force. We collected and analysed data in order to find a relationship between the applied force and the difficulty of the game. This research focuses on the de-hospitalization of subjects in the middle or final stages of their rehabilitation where the new technologies, like Virtual and Augmented Reality, may improve the experience of repetitive exercises.

The results achieved prove that the force applied by the user to hit the virtual sphere with real cardboard cube is related to sphere speed. In a rehabilitation scenario the results could be used to evaluate the improvements analysing the performance history.

Keywords: Rehabilitation · Virtual and Augmented Reality · Brick serious games

1 Introduction

Stroke is a brain attack that occurs when blood flows to an area of brain is cut off, and it represents a leading cause of disability [1]. This situation causes death of brain cells that are deprived of oxygen. Since a stroke involves brain cells, their death has important consequences on the functions of the human body; in particular, functional deficit depends on the brain region affected by dead cells. If the stroke occurs on the right side of the brain, it will affect the left side of the body, including the left side of face. In this case, paralysis on the left side of the body, poor eyesight, quick and inquisitive behavioural style, memory loss can arise too. When the stroke occurs on the left side of the brain, the right side

© Springer International Publishing Switzerland 2016
F. Rossi et al. (Eds.): WIVACE 2015, CCIS 587, pp. 127–136, 2016.
DOI: 10.1007/978-3-319-32695-5_12

of the body will be affected, producing paralysis on the right side of the body, speech disorder, slow and cautious behavioural style, memory loss. In some cases, stroke can occur in brain stem, where, depending on the severity of injury, it can affect both sides of the body and may leave someone in a locked-in state. When this state occurs, the patient is generally unable to speak or achieve any movement below the neck [2].

Rehabilitation of people affected by a stroke is probably one of the most important phases of recovery for many stroke survivors. Rehabilitation can bring subject to his independent life, if properly addressed. To achieve this goal, monitoring EMG (electromyography) signals during stroke rehabilitation therapy can provide valuable insights on neuromuscular system, leading to increased recovery effectiveness.

In this work, we propose a novel approach to forearm rehabilitation that makes use of Myo armband, a low cost gesture controller, to acquire electromyographic signals. Our approach relies on the execution of rehabilitation tasks of a serious game in Augmented Reality (AR) using a simple video see-through setup. This choice reduces equipment and national healthcare system costs, making the system portable and allowing a true home rehabilitation.

The Myo Gesture Control Armband from Thalmic Labs (Fig. 1) is a chain of plastic, or rectangular "pods", where each of them has one medical grade stainless steel EMG (ElectroMyoGraphic) sensor able to detect the electrical pulses of muscles. Myo armband has 8 distinct sEMG (surface EMG) sensors. It communicates via Bluetooth and sends the 8 samples (one per pod) with a frequency of 50 Hz. EMG is referred to as myoelectric activity. Muscle tissue conducts electrical potentials similar to the way nerves do and the name given to these electrical signals is the muscle action potential. Surface EMG (sEMG) is a method of recording the information present in these muscle action potentials [3].

The muscle activity and the EMG signal associated may vary according to the number of Motor Units (MU) recruited and their activation frequency. A MU is the smallest functional unit, which describes the neural control of muscle contraction. During voluntary muscle contraction, two independent parameters modulates the force applied: the first one represents the number of recruited MUs and the second one is MU activation frequency. Considering an experiment, which involves the same muscle activity and the same applied force, is unlikely to observe the same pattern signal. The main parameters that influences EMG signals are:

- tissue kind
- tissue thickness
- user's temperature
- user's physiological state

Due to the non-reproducibility of EMG signals when performing the mentioned experiment, it is appropriate to apply particular smoothing algorithms to filter signals. Using appropriate processing techniques, such as Average Rectified Value (AVR) or Root Mean Square (RMS), it is necessary to evaluate the average signal trend that represents the applied force.

2   Related Works

Various approaches, different from classical therapy, have been used formerly on post-stroke rehabilitation such as virtual reality and other techniques [4,5]. In [6], it is used an active robot exoskeleton to support patients in their rehabilitation phase doing reaching/grasping tasks. Moreover, they have also developed a real time tracking system in order to let the patient interact with real objects. In [7] an arm rehabilitation robot has been studied in order to perform task-oriented repetitive exercises in patients with neurological and orthopaedic lesions, training patients with everyday life activities. In [8] an arm rehabilitation system based on Augmented Reality was developed; they use an haptic sensor to measure the wrist impedance in order to have strength feedbacks, while the AR system is used as hand tracking system. A similar work could be found in [9], where they perform different tasks for different upper art parts (hand, wrist and arm) in order to evaluate different parameters for each task.

In some cases, it is important to migrate the rehabilitation phase to patients' home where it is possible to adapt the actual environment to personal needs after medical staff understood the best living conditions, for example using Virtual Reality (VR) simulations [10,11] and Augmented Reality (AR) using video or optical See-Through Devices [12,13].

We developed a serious game to evaluate the relationship between the force applied by the user and the difficulty level of augmented reality serious game used in rehabilitation tasks. Compared to state of the art our system offers a low cost and portable augmented reality setup.

In the next sections, we describe the serious game, the acquisition protocol for sEMG signals and their processing in order to find the aforementioned relationship.

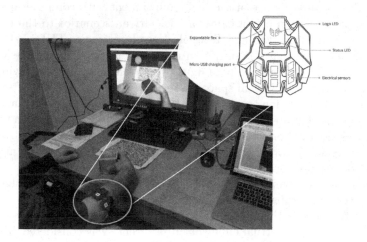

Fig. 1. Serious game setup

3   Methodology

3.1   Experiment Description

The augmented reality serious game designed is inspired by an arcade game developed and published by Atari, Inc. in 1976 and known all over the world. In the game Breakout, shown in Fig. 2, a layer of bricks lines the top third of the screen. A ball travels across the screen, bouncing off the top and side walls of the screen. When a brick is hit, the ball bounces away and the brick is destroyed. The player loses a turn when the ball touches the bottom of the screen. To prevent this from happening, the player has a movable paddle to bounce the ball upward, keeping it in play. The game difficulty is strictly correlated to ball speed.

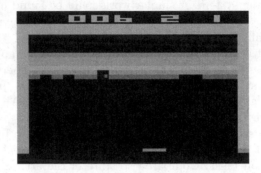

Fig. 2. The original Breakout game by Atari, Inc (1976)

Our serious game incorporates AR technology and tangible interaction techniques to provide an innovative interface to rehabilitation tasks. The system augments the captured image of the real environment with computer-generated graphics to present a variety of game or task-driven scenarios to the user [14]. The tangible interaction (through Tangible User Interface or TUI [15,16]) facilitate to interact with digital information using the physical environment. TUIs have emerged as an alternative paradigm to conventional GUIs allowing users to manipulate objects in virtual space using real, thus tangible, objects.

The designed AR serious game involves the use of two tangible markers, on which 3D objects, that are part of game, are attached. The system, via a webcam, tracks the position and orientation of each tangible markers as it is moved. The used optical tracking system is based on the natural feature tracking algorithm implemented in Vuforia [17]. The "A marker" is associated to the line of bricks and to the game area, while the real TUI "B marker" is associated to the paddle used to hit the moving ball in the game.

In order to simplify the user's movement to achieve certain results, a single target represented by a virtual cube (or cube target) was considered (Fig. 3) in the upper right corner of the "A marker". The initial position of the sphere is in the upper left corner of the scene and, during the game, it moves toward the cube (paddle) that the user grabs in her/his hand (Fig. 4).

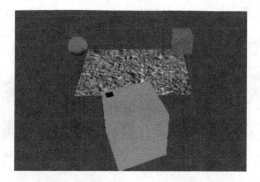

Fig. 3. The initial positions of sphere, paddle and target

(a) First Trajectory   (b) Second Trajectory

Fig. 4. Sphere trajectories

The goal is to hit the cube target with the sphere. Unlike the original game, in this version the user does not move the paddle left-right, but s/he can only rotate the paddle on its vertical axis. This way, we consider only a wrist rotation.

3.2   Testing

System Setup. Starting the experiment, it is necessary to place ARTags in a correct way. The "A marker" must be placed on a table in front of the user, while the "B marker" must be placed between the "A marker" and the user. The user should seat in a way that both, the centre of the marker attached on the cube and the shoulder relative to the forearm under rehabilitation, lie in the same sagittal plane as shown in Fig. 5. The user wears the Myo armband, s/he has to relax the arm to warm up the device. During this phase, the Myo establishes a strong electrical connection with the muscles. It is not necessary to perform the usual calibration. In fact, in this work, the Myo device is used as an array of sensors to record the sEMG signals and not as a gesture controller.

In the initial stage of rehabilitation at home, users have to be trained in order to handle and to move the cardboard cube in the correct way. This is necessary to reduce the noise caused by the tension of other muscles registered involuntarily in the event.

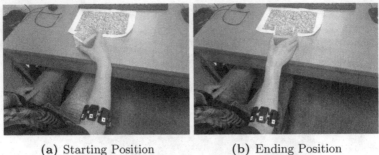

(a) Starting Position   (b) Ending Position

Fig. 5. Handle and movement

Experiment Start. When the application is started, a setting panel is shown, where an operator may insert:

- **IP Server:** the address on which game sends message to start and stop process that saves samples;
- **Name:** the user's name;
- **Initial Speed:** the sphere speed at first attempt;
- **Speed Increase:** the value that is added to sphere speed of the previous attempt to get the sphere speed of the actual attempt;
- **Levels number:** the number of trials in an experiment

An experiment consists of two trials. In particular in the first attempt the sphere has a speed equal to 0.17 m/s while in the second trial it increases to 0.39 m/s. When all the objects are framed on the scene, the sphere, starting from initial position as described previously, moves towards the paddle; the user has to hit the sphere in such a way that it bounces off pointing to the virtual target and s/he has to hold final position until the sphere hits the target or exits the scene. In order to make trials having the same duration, the sphere speed is decreased after hitting the paddle. For each trial we recorded 150 samples.

4   Experimental Results

For this study 5 healthy persons aged between 24 and 30 years (mean 27,2 sd 1,9) were recruited. For each subject the sEMG signal related to the two trials was registered. First of all, collected samples were processed to detect the two sensors with the greatest variance on the measured signal, let name them sensor A and sensor B. In Table 1 the values of variance for each user and each sensor are shown for each attempt.

Then it was performed an amplitude analysis on the samples recorded by the two sensors identified in the previous step. It was calculated the RMS which represents an evaluation of the force applied by the user during the test [18]. Table 2 shows the results. Finally, it was normalized the forces on the value obtained in the trial with higher sphere speed (Table 3).

Table 1. Variance for each user, attempt and sensor

User	Attempt	Variance							
		Sensor 1	Sensor 2	Sensor 3	Sensor 4	Sensor 5	Sensor 6	Sensor 7	Sensor 8
User1	1	15,717	39,954	6,8857	5,2445	1,8161	2,4166	5,2088	44,573
	2	64,438	267,28	52,15	11,376	4,2205	5,0629	40,165	48,774
User2	1	25,791	343,44	24,067	16,43	17,333	4,4695	6,1601	6,0148
	2	33,191	811,82	50,73	26,702	28,311	5,3037	11,19	11,551
User3	1	198,82	857,45	88,023	54,348	38,033	32,275	16,984	23,764
	2	353,08	1102,4	110,56	39,936	35,501	19,564	29,805	23,284
User4	1	77,42	514,69	64,463	15,036	14,129	5,8494	19,85	91,609
	2	133,23	768,49	100,22	20,408	16,62	10,426	46,777	218,01
User5	1	17,587	201,64	121,63	48,898	11,282	4,155	25,702	125,14
	2	77,925	873,64	1148,1	374,18	35,847	12,066	13,237	162,76

Table 2. RMS values for each sensors

User	Attempt	RMS values		
		Sensor A	Sensor B	Average
User1	1	6,3322	6,6696	6,5009
	2	8,0503	16,403	12,22665
User2	1	5,0599	18,735	11,89745
	2	28,445	7,1401	17,79255
User3	1	14,072	29,187	21,6295
	2	18,75	33,284	26,017
User4	1	22,669	9,6737	16,17135
	2	27,614	14,769	21,1915
User5	1	14,164	11,198	12,681
	2	29,573	33,799	31,686

5   Discussion and Results

In order to consider only the sensors activated during the movement, it was calculated samples variance for each sensor. It were chosen the first two sensors on which was registered greatest variance (sensor A and sensor B). Samples registered by these sensors were used to calculate the mean power of the signal which represents an evaluation of force applied by user during experiment. The value considered to compare the force applied in the two trials is an average between mean power of signal registered by sensor A and sensor B.

In the Table 2 it is possible to observe that the applied force is proportional to the sphere speed (Fig. 6).

In a hypothetical rehabilitation scenario a supervisor could set a game level suited to the physiological conditions of the user and then monitor the rehabilitation performance in terms of applied force.

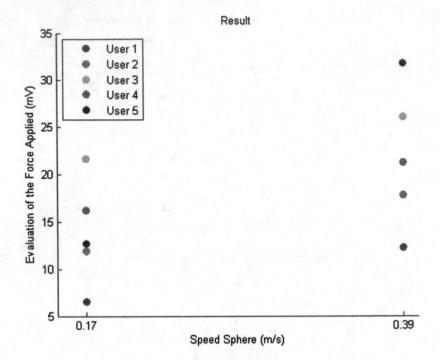

Fig. 6. Result

Table 3. Normalization

User	Attempt	Normalization
User 1	1	0,531699198
	2	1
User 2	1	0,668675935
	2	1
User 3	1	0,831360264
	2	1
User 4	1	0,76310549
	2	1
User 5	1	0,400208294
	2	1

6   Conclusion

In this work, we propose a system for forearm rehabilitation based on an AR serious game using a video see-through device.

We use a low cost and portable setup including a gesture controller device to monitor the rehabilitation task performances recording and processing sEMG signals. The samples are processed to find an evaluation of the applied force during the rehabilitation task related to level of difficulty of the game. In a rehabilitation scenario the results could be used to evaluate the improvements analysing the performance history.

References

1. Goldstein, L.B., Bushnell, C.D., Adams, R.J., Appel, L.J., Braun, L.T., Chaturvedi, S., Creager, M.A., Culebras, A., Eckel, R.H., Hart, R.G., et al.: Guidelines for the primary prevention of stroke a guideline for healthcare professionals from the american heart association/american stroke association. Stroke **42**(2), 517–584 (2011)
2. James, R.: Patterson and Martin Grabois. Locked-in syndrome: a review of 139 cases. Stroke **17**(4), 758–764 (1986)
3. Reaz, M.B.I., Hussain, M.S., Mohd-Yasin, F.: Techniques of emg signal analysis: detection, processing, classification and applications. Biol. Proced. Online **8**(1), 11–35 (2006)
4. Nguyen, K.D., Chen, I., Luo, Z., Yeo, S.H., Duh, H.B.-L., et al.: A wearable sensing system for tracking and monitoring of functional arm movement. IEEE/ASME Trans. Mechatron. **16**(2), 213–220 (2011)
5. Yeh, S.-C., Lee, S.-H., Wang, J.-C., Chen, S., Chen, Y.-T., Yang, Y.-Y., Chen, H.-R., Hung, Y.-P.: Virtual reality for post-stroke shoulder-arm motor rehabilitation: Training system & assessment method. In: 2012 IEEE 14th International Conference on e-Health Networking, Applications and Services (Healthcom), pp. 190–195. IEEE (2012)
6. Loconsole, C., Stroppa, F., Bevilacqua, V., Frisoli, A.: A robust real-time 3D tracking approach for assisted object grasping. In: Auvray, M., Duriez, C. (eds.) EuroHaptics 2014, Part I. LNCS, vol. 8618, pp. 400–408. Springer, Heidelberg (2014)
7. Nef, T., Riener, R.: Armin-design of a novel arm rehabilitation robot. In: 9th International Conference on Rehabilitation Robotics, ICORR 2005, pp. 57–60. IEEE (2005)
8. Khademi, M., Hondori, H.M., Lopes, C.V., Dodakian, L., Cramer, S.C.: Haptic augmented reality to monitor human arm's stiffness in rehabilitation. In: 2012 IEEE EMBS Conference on Biomedical Engineering and Sciences (IECBES), pp. 892–895. IEEE (2012)
9. Hondori, H.M., Khademi, M., Dodakian, L., Cramer, S.C., Lopes, C.V.: A spatial augmented reality rehab system for post-stroke hand rehabilitation. In: MMVR, pp. 279–285 (2013)
10. Bevilacqua, V., Brunetti, A., de Biase, D., Tattoli, G., Santoro, R., Trotta, G.F., Cassano, F., Pantaleo, M., Mastronardi, G., Ivona, F., et al.: A p300 clustering of mild cognitive impairment patients stimulated in an immersive virtual reality scenario. In: Intelligent Computing Theories and Methodologies, pp. 226–236. Springer (2015)
11. Ricci, K., Delussi, M., Montemurno, A., Brunetti, A., de Biase, D., Tattoli, G., Trotta, G.F., Cassano, F., Santoro, R., Pantaleo, M., Mastronardi, G., Ivona, F., Bevilacqua, V., de Tommaso, M.: A p300 clustering of old patients stimulated in an immersive virtual reality scenario with oculus rift. Neuropsychological Trends (2015)

12. Fiorentino, M., Debernardis, S., Uva, A.E., Monno, G.: Augmented reality text style readability with see-through head-mounted displays in industrial context. Presence Teleoperators Virtual Environ. **22**(2), 171–190 (2013)
13. Debernardis, S., Fiorentino, M., Gattullo, M., Monno, G., Uva, A.E.: Text readability in head-worn displays: Color and style optimization in video versus optical see-through devices. IEEE Trans. Vis. Comput. Graph. **20**(1), 125–139 (2014)
14. Burke, J.W., McNeill, M.D.J., Charles, D.K., Morrow, P.J., Crosbie, J.H., McDonough, S.M.: Augmented reality games for upper-limb stroke rehabilitation. In: 2010 Second International Conference on Games and Virtual Worlds for Serious Applications (VS-GAMES), pp. 75–78. IEEE (2010)
15. Fiorentino, M., Uva, A.E., Monno, G., Radkowski, R.: Augmented technical drawings: a novel technique for natural interactive visualization of computer-aided design models. J. Comput. Inf. Sci. Eng. **12**(2), 024503 (2012)
16. Fiorentino, M., Monno, G., Uva, A.E.: Tangible digital master for product lifecycle management in augmented reality. Int. J. Interact. Des. Manufact. (IJIDeM) **3**(2), 121–129 (2009)
17. Vuforia. https://developer.vuforia.com/
18. Milner-Brown, H.S., Stein, R.B.: The relation between the surface electromyogram and muscular force. J. Physiol. **246**(3), 549–569 (1975)

Systems Chemistry and Biology

Ab-initio Investigation of Unexpected Aspects of Hydroxylation of Diketopiperazines by Reaction with Dioxiranes

Cosimo Annese[1,2], Lucia D'Accolti[1,2], Caterina Fusco[2], and Fulvio Ciriaco[1(✉)]

[1] Dip. di Chimica, Università degli studi di Bari, via Orabona 4, 70126 Bari, Italy
fulvio.ciriaco@uniba.it
[2] CNR ICCOM-Bari, UOS Bari, via Orabona 4, 70126 Bari, Italy

Abstract. In an explorative study of the opportunities for synthesis provided by oxidation of natural aminoacidic compound with methyl-(trifluoromethyl)dioxirane, we noticed that oxidation of the cyclic compound, when successful, leads to the ?-hydroxy compound, notwithstanding the presence of the isopropyl groups that are usually easy targets for this reaction, as verified with the homologous acyclic compound. We therefore initiated an ab-initio study of the reactions aimed at determining the role of the ring and explaining the reactivity differences of the cis and trans configurations. Consistently with recent literature, we confirm the fundamental role of an adduct configuration in which the dioxirane O O bond is largely divaricated and electron pairing is broken, often denoted in the literature as diradicaloid.

Keywords: DFT · B3LYP · Transition state · Openshell · Diketopiperazine · Dioxirane · Hydroxylation · PES · Radical · Diradical

1 Introduction

Oxidation of amino acid residues of peptides and proteins by reactive oxygen species (ROS) such as singlet oxygen, superoxide, or hydroxyl radical is implicated in a number of pathological disorders, as well as in the progression of aging. A number of studies have been devoted to the characterization of products from reaction of alkyl amino acid residues with ROS [4], metal-based oxidants [6], as well as biomimetic oxidation systems [2]. The common feature of all these oxidative transformations is the selective oxidation of amino acid residues, providing α-hydroxyl amino acids or products of backbone oxidative cleavage [1,7]. By a way of contrast, we and other authors have shown that reaction of alkyl amino acid residues with methyl(trifluoromethyl)dioxirane (TFDO) results in the hydroxylation of the aliphatic side chain, so that Val and Leu residues give β-hydroxyvaline and γ-hydroxyleucine, respectively, while Gly and Ala residues do not undergo oxidation.

Rauk et al. studied the γ- vs. α-CH dioxirane oxidation of Leu by means of B3LYP DFT computations. In disagreement with experiment, a slightly lower

© Springer International Publishing Switzerland 2016
F. Rossi et al. (Eds.): WIVACE 2015, CCIS 587, pp. 139–145, 2016.
DOI: 10.1007/978-3-319-32695-5_13

(3 kJ/mol) activation barrier was predicted for dioxirane oxidation of the α-CH of N-formylglycine amide in the gas phase; a better agreement with experimental data was obtained by taking into account the influence of the solvent CH_2Cl_2 using the IPCM model. All these observations pointed to an electrophilic dioxirane oxidation of the C-H bond via a polar transition state, which also has partial diradical character [8].

Herein, we expand the substrate scope to cyclic dipeptides, such as cyclo(Val-Val). These small-ring cyclic peptides are conformationally constrained heterocyclic scaffolds and are important in drug discovery because they have a rigid backbone, which can mimic a preferential peptide conformation and contain constrained amino acids embedded within their structures without the unwanted physical and metabolite properties of peptides. Unexpectedly, we found that TFDO oxidation of cyclo(L-Val-L-Val) and of cyclo(L-Val-D-Val) exclusively occurs at the α-CH, providing the corresponding α-hydroxy derivatives. These findings prompted us to theoretically assess the characteristics of the O-transfer from TFDO to the α- and β-C-H bond of cyclo(L-Val-L-Val) and cyclo(L-Val-D-Val), by taking into account recent developments in the theoretical description of the C-H bond oxidation by dioxiranes [11].

2   Experimental Setup and Results

Cyclo(L-Val-L-Val) (**a**) and cyclo(D-Val-L-Val) (**b**) were obtained by cyclization of Boc-L-Val-L-ValOCH$_3$ and Boc-D-Val-L-Val-OCH$_3$, respectively [9].

Oxidations of substrates **a** and **b** were performed with solutions of TFDO in the parent ketone, and were carried out in a mixture CH_2Cl_2/t-BuOH 3:1, which represented the best choice in terms of substrate solubility. Reaction yields and conversions were determined by HPLC analysis of the reaction mixture, while product isolation and purification entailed reversed-phase preparative HPLC. Typical results and reaction conditions are collected in Fig. 1.

As shown in Fig. 1, the oxidation of substrate **a** provides cyclo(L-α-hydroxy-Val-L-Val) (**a-OH**) as the main product (95 %), along with 5 % of the bis(hydroxy) derivative **a-2OH**, notwithstanding the presence of the β-CH groups, easily oxidizable with dioxiranes, as already shown in the case of linear N-Ac-L-Val-OCH$_3$ and N-acetyl-peptides containing Val residues [1,7]. Oxidation of cyclo(D-Val-L-Val) (**b**, Fig. 1) attains low conversion under the same conditions, although the α-regioselectivity is still totally maintained, as indicated by the exclusive formation of the hydroxy product **b-OH**. The ^{1}H- and ^{13}C NMR spectra of reaction products were in excellent agreement with the structures proposed. In particular, it is seen that the C$_2$-symmetry of substrate **a** is maintained in the product **a-2OH**. For instance, the ^{13}C NMR spectrum of **a-2OH** shows a signal at 82.95 ppm, due to the resonance of the two αC-OH, while the two β-CH carbons show a signal at 35.25 ppm.

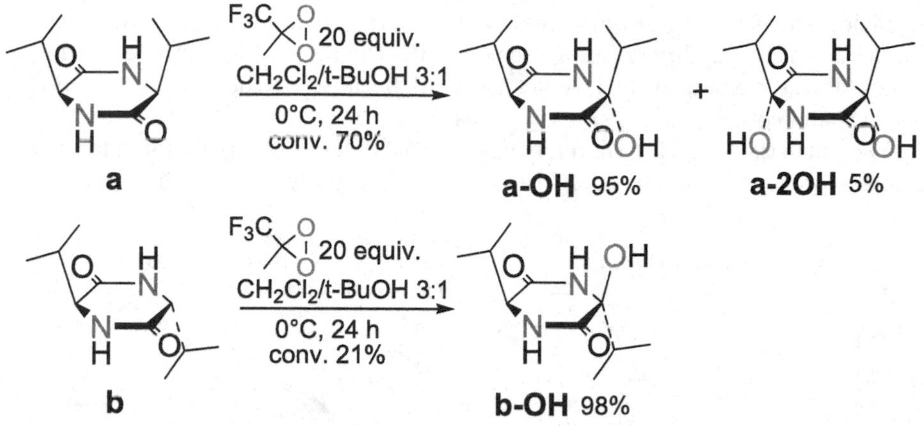

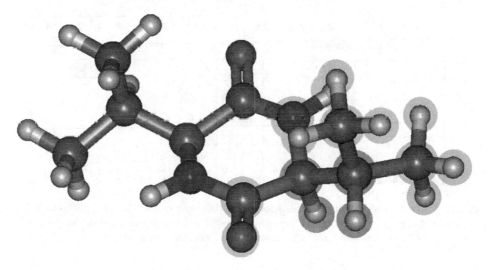

Fig. 1. Substrates, products, reaction conditions and yields.

Fig. 2. Substrate cyclo(L-Val-L-Val) geometry, atoms treated with PC-2 basis set are highlighted; the orange highlight for atoms directly involved in the hydroxylation

3   Computational Methods and Details

All computations were essentially DFT (B3LYP) constrained geometry optimizations and where conduced by means of the suite for electronic structure ab-initio studies NWChem [10].

To cut down on the resources needed for the large substrates, we made use of the hierarchical polarization consistent basis set from Jensen [5]: PC-2 for the region of interest, i.e. the atoms directly involved in the reaction and up to the second nearest neighbours from these, also indicated with a yellow halo in Fig. 2 and PC-1 for the rest of the system.

Calculations were performed testing both attack directly at the carbon atom and attack at the hydrogen. In agreement with the recent literature [3,8], direct attack at the carbon atom is not successful because of the high involved TS energies and corresponding side reactions.

We will in the following therefore describe only the results pertaining to attack at the hydrogen atom. The energetics of the reaction are depicted in Fig. 3.

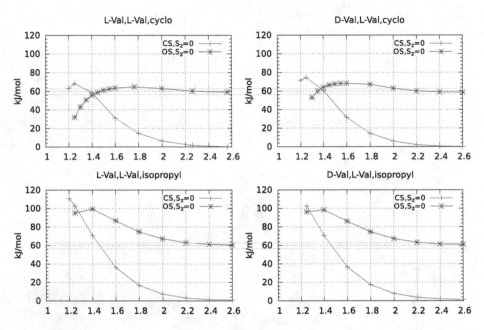

Fig. 3. Paths to the transition states for the 4 different cases: the two different substrates and the two different hydroxylation sites. The abscissa represent the distance between the hydrogen atom and the nearest dioxirane oxygen atom.

The transition state for hydroxylation of β-CH is higher compared to that of reaction at the site in the ring by almost $40\,\mathrm{kJ/mol}$.

Each of the four plots presents two curves: one for the closed shell lowest state and one for the relevant openshell system with $S_z = 0$.

There is a small difference of almost $10\,\mathrm{kJ/mol}$ between the cyclo(L-Val,L-Val) and the cyclo(D-Val,L-Val) substrates, in fact in agreement with the experiments described above, however the difference is rather small compared to the precision of the present calculations. The same calculations, performed with a COSMO simulation of the solvent with dielectric constant 8.3, but limited to the critical points do not differ importantly.

The approach to the hydrogen atom is only the first step of the reaction, according to the literature [3,8] the reaction progress is easier from the open shell state, through a further TS characterized by shorter CO length and larger COH angle.

In both cases, the closed shell and open shell curves intersect at ≈ 1.4 Å. The geometries are actually different for the two curves and the system needs some more energy to jump from one curve to another.

Figure 4 illustrates the geometry of the cyclo(L-Val,L-Val) dioxirane adduct at 1.40 Å and the spin density of the open shell state. Meaningful distances and angles are reported in Table 1.

The O-O axis is practically collinear to the C-H bond for the closed shell state and perpendicular for the open shell one. Also, the O-O distance is much larger for the latter than for the former. The C-H bond for the open shell state is long and hence weakened. The spin equidensity surface at 0.02 e^-/bohr^3 is evocative of the incipient formation of a 1 electron second C_3-O_2 bond and of a 1 electron bond between C_1 and O_1.

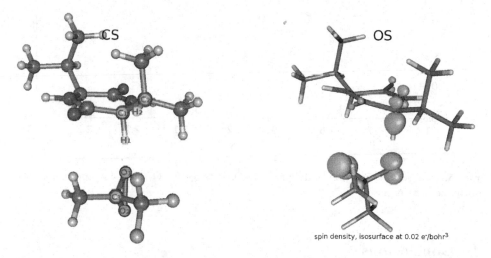

spin density, isosurface at 0.02 e'/bohr^3

Fig. 4. CS: geometry of the adduct at 1.40 Å on the closed shell PES; OS: geometry and spin density of the adduct at 1.40 Å on the open shell PES.

Table 1. Main geometric parameters for substrate-dioxirane adduct at 1.4 Å. Distances in Å, angles in degrees

	O_1-O_2	C_1-H	C_1-H-O_1	H-O_1-O_2	C_3-O_1	C_3-O_2
Closed shell	1.63	1.15	172.3	175.4	1.40	1.37
Open shell	2.31	1.26	169.7	90.8	1.37	1.35

Figure 5 summarizes our attempt to characterize the transition state between the closed and open shell PESes when O-H is 1.4 Å, using the O-O distance as main component of the reaction coordinate. However, when the CS and OS

curves cross in the plot, at about O-O = 1.97 Å, their geometry is still far from a
TS, as assessed by the calculation of a closed shell state at the geometry of OS.
The plot assesses that the TS energy is somewhere between 20 and 140 kJ/mol,
that is however too much undetermined for any reasonable use.

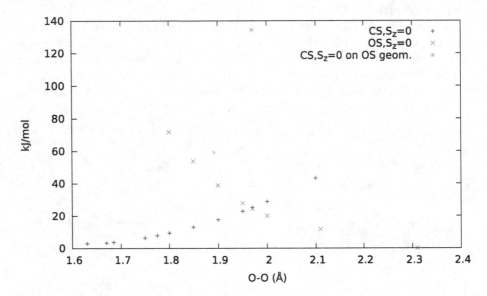

Fig. 5. Energy as a function of dioxirane O-O distance for O-H = 1.4 Å and otherwise
optimized geometry.

4   Conclusions

The recent literature has much disambiguated the reaction of hydroxylation of
C-H bonds with dioxiranes, mostly contradicting previous beliefs. The present
paper confirms the basic mechanism of the reaction in its first stage, i.e.
attack of dioxirane oxygen to the hydrogen atom even for the case of diketopiper-
azines [11]. We have also conducted studies about the successive stages of the
reaction, but the presence of many interactions with the complicated substrate
makes very difficult to rationalize and present the results.

Within the limits of the present calculations, this study confirms the experimental observations about the reaction regioselectivity.

The probable multi-reference nature of the transition state of this reaction
suggests that multi-reference methods should be used for a more accurate representation of the mechanism.

Acknowledgments. The research was supported by CNR ICCOM and Regione Puglia MIUR PON Ricerca e Competitività 20072013 Avviso 254/Ric. del 18/05/2011, Project PONa3 00369 Laboratorio SISTEMA.

Commercial Caroat® triple salt (2KHSO$_5$ · KHSO$_4$ · K$_2$SO$_4$) was kindly supplied by Peroxide-Chemie, (Degussa, Germany).

We acknowledge the CINECA award under the ISCRA initiative, class C, project dichetox, for the availability of high performance computing resources and support.

References

1. Annese, C., Fanizza, I., Càlvano, C., D'Accolti, L., Fusco, C., Curci, R., Williard, P.: Selective synthesis of hydroxy analogues of valinomycin using dioxiranes. Org. Lett. **13**, 5096–5099 (2011)
2. Ekkati, A., Kodanko, J.: Targeting peptides with an iron-based oxidant: cleavage of the amino acid backbone and oxidation of side chains. J. Am. Chem. Soc. **129**, 12390–12391 (2007)
3. Freccero, M., Gandolfi, R., Sarzi-Amadè, M., Rastelli, A.: Novel pathways for oxygen insertion into unactivated C-H bonds by dioxiranes. Transition structures for stepwise routes via radical pairs and comparison with the concerted pathway. J. Org. Chem. **68**, 811–823 (2003)
4. Hawkins, C., Morgan, P., Davies, M.: Quantification of protein modification by oxidants. Free Radical Biol. Med. **46**, 965–968 (2009)
5. Jensen, F.: Polarization consistent basis sets: principles. J. Chem. Phys. **115**, 1 (2015)
6. Murahashi, S., Mitani, A., Kitao, K.: Ruthenium-catalyzed glycine-selective oxidative backbone modification of peptides. Tetrahedron Lett. **41**, 10245–10249 (2000)
7. Rella, M., Williard, P.: Oxidation of peptides by methyl(trifluoromethyl)dioxirane: the protecting group matters. J. Org. Chem. **72**, 525–531 (2007)
8. Shustov, G., Rauk, A.: Mechanism of dioxirane oxidation of CH bonds: application to homo- and heterosubstituted alkanes as a model of the oxidation of peptides. J. Org. Chem. **63**, 5413 (1998)
9. Tullberg, M., Grotli, M., Luthman, K.: Efficient microwave assisted syntheses of 2,5-diketopiperazines in aqueous media. Tetrahedron **62**, 7484–7491 (2006)
10. Valiev, M., Bylaska, E., Govind, N., Kowalski, K., Straatsma, T., van Dam, H., Wang, D., Nieplocha, J., Apra, E., Windus, T., de Jong, W.: NWChem: a comprehensive and scalable open-source solution for large scale molecular simulations. Comput. Phys. Commun. **181**, 1477 (2010)
11. Zou, L., Paton, R., Schenmoser, A., Newhouse, T., Baran, P., Houk, K.: Enhancedreactivity in dioxirane CH oxidations via strain release: a computational and experimental study. J. Org. Chem. **78**, 4037 (2013)

On Fine Stochastic Simulations
of Liposome-Encapsulated PUREsystem™

Lorenzo Calviello[1], Lorenzo Lazzerini-Ospri[2], and Roberto Marangoni[3,4(✉)]

[1] Max Delbrück Center for Systems Biology, Berlin, Germany
calviello.1@gmail.com
[2] Johns Hopkins University, Baltimore, MD, USA
ospri@jhmi.edu
[3] Dipartimento di Biologia, Università di Pisa, Pisa, Italy
roberto.marangoni@unipi.it
[4] Istituto di Biofisica del CNR, Pisa, Italy

Abstract. The PURESystem™ (for short: PS) is a defined set of about 80 different macromolecular species which can perform protein synthesis starting from a coding DNA. To understand the processes that take place inside a liposome with entrapped PS, several simulation approaches, of either a deterministic or stochastic nature, have been proposed in the literature. To correctly describe some peculiar phenomena that are observed only in very small liposomes (such as power-law distribution of solutes and supercrowding effect), a stochastic approach seems necessary, due to the very small average number of molecules contained in these liposomes. Here we recall the results reported in other works published by us and by other Authors, discussing the importance of a stochastic simulation approach and of a fine description of the system: both these aspects, in fact, were not properly acknowledged in such previous papers.

Keywords: PURESystem™ · Liposome · Stochasticity · Supercrowding effect · Power-law distribution

1 Introduction

Among the different cell-free protein synthesis systems, the PUREsystem™ (Protein synthesis Using Recombinant Elements; PS for short hereafter), developed in 2001 by Ueda and coworkers [1] and commercially available, is one of the most known and employed, in particular in synthetic biology studies. Despite the relatively small number of chemical species contained in it (mainly DNA, ribosomes, a tRNA mixture, 36 purified *E. coli* enzymes, plus several small metabolites, for a total of about 80 different macromolecular species), it shows excellent performance in producing proteins starting from coding DNA [2]. Its description and preparation meet the standardization requirements of synthetic biology, it is listed in the Registry of Standard Biological Parts [3].

The studies of biochemical pathways *in vitro* proceeded in parallel to those performed *in vivo*, since *in vitro* approaches allow to test or measure quantities,

© Springer International Publishing Switzerland 2016
F. Rossi et al. (Eds.): WIVACE 2015, CCIS 587, pp. 146–158, 2016.
DOI: 10.1007/978-3-319-32695-5_14

kinetic parameters and other variables, that are not easily accessible *in vivo*. Together with the experiments, the necessity of a model able to describe the principal pathways lead several Authors to propose different strategies, focused on different pathways and using different detail level and different formal approaches. From early works where cell-free synthesis has been used to synthetize different isoforms of a protein [4], to more recent models where protein synthesis is studied *in vivo* or *in vitro*, by either "standard" differential-equation based models [5] or stochastic approaches to understand the role of the noise in the main biochemical patways [6,7], trying to investigate emergent features due to the biochemical network complexity [8].

Apart from these general approaches, the PS has been successfully applied *in lipo*[1], allowing the possibility to monitor and study gene expression in small volumes.

For instance, an interesting study used the PS to study membrane proteins and pore-complex assembly using a novel approach called *liposome display*, where different DNA constructs are used as a template for in vitro protein production and assembly on the liposome membrane [9]. Other numerous applications of the compartmentalized PS were focused on understanding gene expression inside liposomes, such as in the presence of different membrane lipid composition [10], with different preparation methods [11], and in different-sized vesicles [12].

Understanding gene expression in small compartments is of crucial importance for defining the so-called *minimal cell* [10,13,14] a minimal entity able to display the fundamental properties of living system; *the PS is the most used system for the construction of semi-synthetic minimal cells* [13,14].

Indeed, semi-synthetic minimal cells often consist of a liposome enclosing defined biochemical pathways, specifically pathways capable to synthesize proteins that could, in principle, close the circle towards a complete self-sustainability of the minimal cell: in other words, the different components of the minimal cells can cooperate to restore themselves. As of today, only a few papers have been published dealing with theoretical simulations, at different detail level, of PS behavior and time course [7,15–17], and only three deal with PS entrapped in liposomes [18–20]. This last condition is very interesting since the smaller the liposome, the lower the probability that a large amount of molecules of each constituent chemical species be entrapped in the volume of the liposome. This implies that a standard simulation of PS based on deterministic ODE formalism is no longer suitable to describe a biochemical system beyond the deterministic limit. If even a single chemical species is present with only a few molecules, the behavior of all the PS is turned to stochastic.

To describe a stochastic chemical system, the most often used approach is to employ Gillespie algorithm [21]: this is an algorithm derived from collision theory that operates by selecting, by means of two suitably generated random numbers, which reaction r to execute at the current step, and the waiting time

---

[1] By the term *in lipo* we want to indicate a particular *in vitro* procedure, where the biochemical system studied is entrapped into compartments, the walls of which are made by lipid molecules.

τ for the next reaction. The probability distribution of r is uniform over the propensity of the reactions, while the τ distribution is an exponential decay (i.e., the reaction system is described as a pure Markovian process). Gillespies algorithm has been proved to be *correct*: for large numbers of molecules and for long times, the time courses described by the algorithm are identical to those described by the Chemical Master Equation for the continuous case. Because of this characteristic, Gillespies algorithm (with several variants and improvements) has been applied in a very large number of stochastic simulations of chemical, and even biochemical, systems [22].

2   Describing the PS in a Stochastic Simulator

There are several problems to address in order to have a detailed description of the PS in a Gillespies algorithm-based simulator. These problems arise from the not negligible differences between a common chemical system as hypothesized by Gillespie, and a cell-free system, which hosts several complex multi-step processes. For instance, two major problems are represented by the description of transcription and translation: as they imply the presence of a nucleic acid (either DNA or RNA) with an extensive reaction machinery (either polymerases or ribosomes) bound to it. These complexes are spaced from one another by a certain minimum distance, and slide on the nucleic acid molecules moving one step only when the complex ahead has itself moved.

Even this description is very simplified; at any rate, it is incorrect to represent this as a standard stoichiometric system. The concept of sequentially ordered movements is not contemplated in Gillespies approach, and there is no natural way to describe it. A possible solution to bypass these obstacles is to use some *dummy* chemical species, which do not exist in reality, but which can be used to the required effect (the term dummy is introduced by analogy with computer programming, where dummy variables indicate placeholders variables). Therefore, we simulated the attachment and movement of polymerases and ribosomes through a series of dummy chemical species simulating binding sites and movement from a nucleic acid segment to the next.

Adding to the complexity of this scenario, we have to note that translation and transcription are simultaneous events in cell-free PS, and translation starts immediately after the transcription of the first RNA segment, while polymerases are still working on the DNA molecule. This situation is close to that of prokaryotic cells, where translation and transcription take place both in the cytoplasm. As a consequence, the dummy species and the real reactions for both transcription and translation should be arranged to work together.

In the following, we would like to explain in detail the structure of our stochastic description of liposome-entrapped PS.

3 Characteristics and Properties of a Stochastically Simulated PS

3.1 Principles and General Remarks About the Simulator Used (QDC)

In our case study, the PS is used to synthesize Green Fluorescent Protein (GFP): the DNA input molecules code for the standard GFP sequence. Accordingly with Gillespies approach, all the described chemical species can interact independently of each other, and the solutes entrapped inside the liposome, even in the case of dummy species, are supposed to be well-stirred.

We simulated the system by using QDC, a lab-made simulator, whose core is based on Gillespies direct method, with extensions allowing the user to simulate a metabolic experiment, rather than an isolated metabolic system. A detailed survey of QDC is presented in the literature [23]. Here we briefly recall only the most important concepts. QDC can simulate a metabolic experiment since it implements control functions that allow the user to exert some control on the metabolic system during the simulation. In particular, QDC allows: to simulate the addition of molecules at a given time; to simulate continuous feeding or leaking of some species at a set time rate; to specify the so-called "immediate" reactions (reactions with theoretically infinite propensity, that take place immediately after their stoichiometric conditions are met). In our experiments, we have extensively used immediate reactions to describe the advancement of polymerases and ribosomes on their respective nucleic acids.

QDC outputs three main result files: the time courses of all the chemical species present in the system; the propensity time course for each reaction; the number of times that each reaction has been executed. By analyzing and comparing all these three outputs one can extract information about the status of the system and detect possible biases that might have perturbed the simulation results (e.g., a possible *stiffness* of the system).

3.2 Transcription and Translation Processes

The GFP DNA sequence was divided (according to its length) into 80-bp-long segments, each of which is treated as a distinct *chemical species*, and named, for instance, *DNA1, DNA2, DNA3,....* As a consequence, the polymerization process is represented by several reactions: their core module includes a second-order reaction for nucleotide binding, and a first-order reaction for nucleotide incorporation, which returns the polymerase molecule (which, in turn, can bind to another nucleotide) and a dummy product designed to track the number of nucleotides incorporated in the RNA molecule (these will be named *RNA1, RNA2, RNA3, ...*

Some immediate reactions determine the transition to the next step, ensuring the following conditions are met: (a) an adjacent DNA site is available, (b) the correct number of nucleotides has been added to the RNA sequence, (c) the corresponding RNA sequence is produced, (d) the previously occupied DNA site is released. Here is an example for transcription at the fourth step (GTP and ATP are considered):

```
1000000, T7ELGT4 + GTP > T7pregEL4
28, T7pregEL4 > T7ELAT4 + Pi + gtr4
1000000, T7ELAT4 + ATP > T7preaEL4
28, T7preaEL4 > T7ELGT4 + Pi + atr4
-, 20 gtr4 + 20 atr4 + T7ELGT4 + DNA5 > T7ELGT5 + RNA4 + DNA3
```
(These few lines are extracted from the QDC's input file for transcription/translation processes simulation, and show the fourth step in Polymerase activity. In this case, DNA5, RNA4 and DNA3 are dummy species. The numeric coefficients before the reactions are the relative stochastic coefficients, while the minus sign "-" indicates that the corresponding reaction belongs to immediate type. For a detailed explanation of this syntax see [23])

This reaction routine was repeated different times ensuring the correct succession of molecular states; when only one DNA molecule is available, the polymerases advance one by one, separated by at least one DNA site between each other (elongating RNA polymerases are separated from each other by at least 80 bp [24]). The same strategy was used to describe the translation reactions.

An elongating ribosome (eR2) binds the complex carrying the aminoacid (EFaRGTP), after which it moves to the next codon, aided by the elongation factor EFg charged with GTP (EFgGTP); this translocation reaction yields an additional product used to regulate the progression to the next state.

```
100000000, eR3 + EFaRGTP > EXpre3
79, EXpre3 > eR3 + EFaRGTP
207, EXpre3 > EX3
3.45, EX3 > EXpre3
100, EX3 > EX3GDP + Pi
638, EX3GDP > eRa3GDP
15, eRa3GDP > eRa3 + EFtuGDP
20, eRa3 > eRa3tRNA
150000000, eRa3tRNA + EFgGTP > EXbpre3
140, EXbpre3 > eRa3tRNA + EFgGTP
250, EXbpre3 > EXb3 + EFg + GDP +Pi
20, EXb3 > eR3 + tRNA + TRANSL3
-, 27 TRANSL3 + RNA4 + PEPT2 + eR3 > eR4 + PEPT3 + RNA2
```
(This is another extract from the QDC's input file for transcription/translation processes simulation, and show the fourth step in Ribosome activity. In this case, TRANSL3, RNA2, RNA4, RNA4, PEPT3, PEPT2, and others, are dummy species. The numeric coefficients before the reactions are the relative stochastic coefficients, while the minus sign "-" indicates that the corresponding reaction belongs to immediate type. For a detailed explanation of this syntax see [23])

After a fixed number of translocation steps (the minimal space between two elongating ribosomes is, as with polymerases, 80 nt $\sim$ 20 codons [25]) an immediate reaction occurs in a similar fashion as seen for transcription:

1. the correct amount of aminoacids are incorporated, and thus consumed;
2. the next free RNA site is occupied, and
3. the previous one is therefore freed;
4. an entity named PEPT is also produced, tracking the length of the peptide sequence produced so far:

For example, if 4 species named PEPT3 are present in a certain time of the simulation, this means that there are 4 peptides, still bound to the ribosomes, with a length spanning from $27 \times 3 = 81$ to $(27 \times 4) - 1 = 107$ aminoacids. This "segmental" partition of DNA and RNA, with the discrete and stepwise occupation by the transcription and translation molecular machineries respectively, is schematically represented in Fig. 1.

3.3 The Overall PS

As this system focuses on the synthesis of GFP starting from its DNA and other basic molecules (nucleotides, aminoacids, etc.), it hosts only a few reactions other than translation and transcription, namely the initialization of both processes and the eventual ribosome inactivation (which takes place 3 h after the start of protein synthesis, see [17]).

4   Results: Simulating *in-lipo* PS at Nanoscale Range

4.1   Poisson vs. Power-Law: Which Distribution for Solute Entrapment in Nano-Vesicles?[2]

The process of solute entrapment during the formation of liposomes is of special interest, as it affects the distribution of molecules inside them - a relevant issue for studies on the origin of life. Theoretically, when no interactions are supposed to exist between the chemical species to be entrapped, or between these and the nascent lipid bilayer, a standard Poisson process well describes the entrapment mechanism. Recent experimental findings, however, show that, for small liposomes (100 nm diameter), the distribution of entrapped molecules is best described by a power-law function [26], where supercrowded liposomes (i.e., liposomes showing a very high inner solute concentration) are present in a very small but not negligible number. This is a matter of great consequence, as the two random processes generate two completely different scenarios.

We used QDC to simulate a GFP-synthesizing PS encapsulated into liposomes, with the solute partition inside the vesicles obtained by the two entrapment models: a pure Poisson process or a power law. The protein synthesis *in*

---

[2] Reviewed after [19].

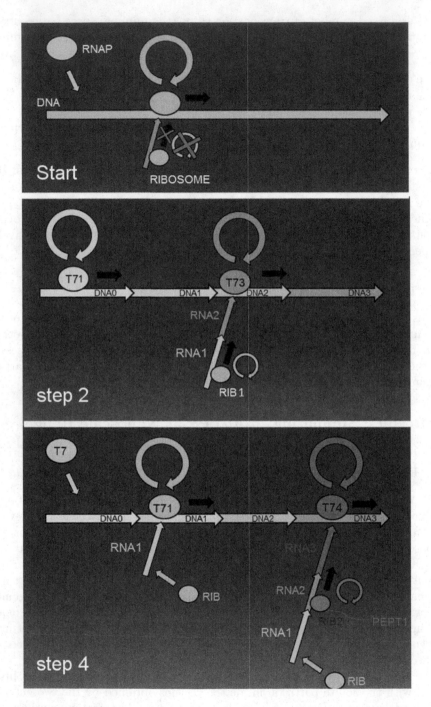

Fig. 1. A schematic representation of the stepwise transcription and co-translation processes (redrawn from [20]).

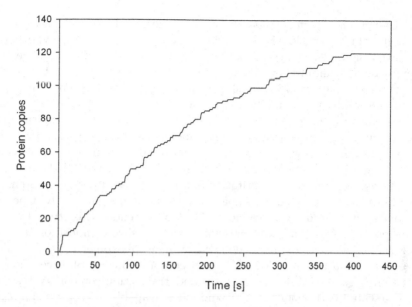

Fig. 2. Time course of the GFP synthesis in a 10^{-17} L vesicle, populated accordingly to a power-law distribution. In the abscissa the time (in s), in the ordinate the number of GFP molecules synthesized (redrawn from [19]).

lipo was studied in both cases to highlight experimental observables that could be measured to test which model best fits the real entrapment process.

The time course of protein synthesis within each kind of vesicle was then simulated, as a function of vesicle size. Our study can predict translation yield in a population of small liposomes down to the attoliter (10^{-18} L) range. Our results show that the efficiency of protein synthesis peaks at approximately $3 \cdot 10^{-16}$ L (840 nm diam.) with a Poisson distribution of solutes, while a relative optimum is found at around 10^{-17} L (275 nm diam.) for the power-law statistics. Figure 2 shows the time course of GFP synthesis in a vesicle of 10^{-17} L volume: the protein copies accumulate over time, until the depletion of inner reagents stops the generation of new copies.

Our simulation clearly shows that the wet-lab measurement of an effective protein synthesis at smaller volumes than 10^{-17} L would rule out, according to our models, a Poisson distribution of solutes, and thus would indirectly support that a supercrowding effect took place. This suggestion fits well with a discussion found in [27].

4.2 How Is Intra-vesicle Solute Composition Driving Gene Expression from DNA to Protein?[3]

The next step was to determine whether all chemical constituents of the PS are entrapped in liposomes according to a power law (then, giving rise to a

---

[3] Reviewed after [20].

supercrowding effect) independently of each other. Independence (or lack thereof) in solute partition is a foundational feature which we hypothesized would have a significant effect on the final state of the system (e.g. efficiency of gene expression). Finding key end-state parameters dependent on the original entrapment dynamics would also offer an obvious pathway to empirical testability. In this study, the simulations were firstly performed for large liposomes (2.67 μm diameter) entrapping the PS to synthesize GFP. By varying the initial concentrations of the three main classes of molecules involved in the PS (DNA, enzymes, consumables), we were able to stochastically simulate the time-course of GFP production. A mathematical fitting of the simulated GFP production curves allowed us to extract quantitative parameters describing the protein production kinetics; as expected, these parameters resulted significantly dependent on the initial inner solute compositions. Then we extended this study to small-volume liposomes (575 nm diameter), where intra-vesicle composition is harder to infer due to the expected anomalous entrapment phenomena.

In our *in silico* perturbation of the PS composition, we observed how the competition between DNA transcription and translation for energy resources (here represented by the sum of ATP and GTP available molecules) shapes the protein production dynamics, in terms of total protein yield and its production kinetics. Figure 3 offers a global survey of these results: it shows, for each class of initial DNA inner concentration, the resulting global production in GFP and the corresponding use of energy molecules by the different biochemical processes.

A mathematical formalization of the PS reaction network allowed us to focus on different aspects of *in lipo* protein production, and a discussion about our computational strategy together with its scientific relevance is presented in the following sections.

5   On the Relevance of a Detailed Stochastic PS *in Silico* Model

Different *in silico* approaches have been tackled for the description of the complex system of reactions encompassing the PS. Despite the large number of biochemical species, the interactions between the single molecules are well-known. Moreover, higher-order interactions between different molecular classes seem to have a modest impact on the overall protein production [29]. Therefore, it is reasonable to approach a comprehensive description of the PS by modeling every reactant as a single entity.

A deterministic formulation of transcription and translation processes can be a useful representation of the PS [17] under the assumption of a continuous, homogeneous distribution of solutes. When dealing with compartmentalized biochemical networks, the intrinsic stochasticity of molecular reactions plays an important role in the observed protein production. A very recent study [30] highlighted the importance of stochasticity in cell-sized lipid compartments, showing how fluctuations in the number of DNA molecules is one of the main responsible of protein production noise using the PS. As volume decreases, well-described

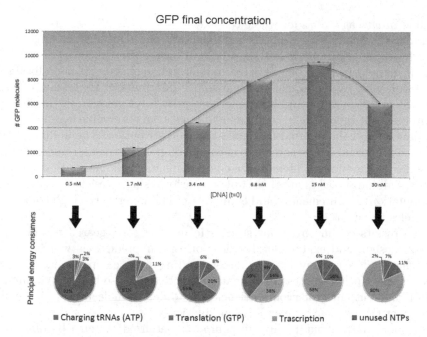

Fig. 3. The graph shows the final GFP production obtained (ordinate) according to the initial DNA inner concentration (abscissa). The pie charts placed in correspondence of the abscissa summarize the global energy molecules use for all the case studied (redrawn from [20]).

anomalous entrapment phenomena [26] create a drastic vesicle-to-vesicle variability in terms of protein production. Moreover, the extremely low number of molecules that eventually can govern a cellular process poses a great challenge in modeling the PS biochemical network. A stochastic representation of the PS deals with those limitations, allowing researchers to focus on single components of the network, with the possibility to focus on nanoscale-sized compartments. This creates the ability to further test different assumptions of the encapsulation efficiencies for different groups of reactant, which, by differentially interacting with the lipidic bilayer, can differentially drive protein production in a size-dependent manner.

Moreover, quantifying the extent to which the intrinsic vesicle-to-vesicle variability influences the protein production in small compartments is of great relevance to discriminate between different components of variability in nano-scaled compartmentalized gene expression: one coming from the random collision between molecules, and the other coming from different internal vesicles composition.

6   Conclusions and Future Directions

Cell-free technologies like the *in lipo*-PS allow us to monitor and study gene expression kinetics under controlled, known conditions, whereas this is not possible *in vivo* as well as for whole cell-extracts (e.g. *E. coli* extracts) also used in similar synthetic biology approaches. Such information is of great value for the most diverse applications, in both applied and basic research [30,31].

Our main focus is the quantitative understanding of compartmentalized gene expression with respect to the internal solute distribution. By testing different entrapment mechanisms and their effect on protein production in the different PS processes, it is possible to create a link between an observable feature (protein production) and the internal liposomal solute distribution. The ability to infer the precise internal vesicle content is of course of great interest for understanding the different kinetics in different size compartments. This means that the ideal experimental approaches are those able to perform single vesicle measurements, as population measurements could be unable to characterize the high complexity of vesicles behavior.

The results coming from our stochastic modeling are geared to aid experimental design, and experimental validation of our predictions will be needed to further explore and improve our understanding of the compartmentalized PS. A better description of liposome formation kinetics will also help the community to test different encapsulation efficiencies for different molecules, which will in turn affect the internal gene expression kinetics.

As more data about PS use in compartmentalized systems become available, our interpretation of *in silico* experiments will also improve, allowing us to further tackle the problem of understanding nano-scaled protein-producing liposomes.

Acknowledgments. The Authors will thank Pasquale Stano and Fabio Mavelli for the valuable discussions during the preparation of this review.

References

1. Shimizu, Y., Inoue, A., Tomari, Y., Suzuki, T., Yokogawa, T., Nishikawa, K., Ueda, T.: Cell-free translation reconstituted with purified components. Nat. Biotechnol. **19**, 751–755 (2001)
2. Shimizu, Y., Kanamori, T., Ueda, T.: Protein synthesis by pure translation systems. Methods **36**, 299–304 (2005)
3. http://parts.igem.org/Chassis/Cell-Free_Systems. Accessed 14 December 2015
4. Giudice, L.C., Chaiken, I.M.: Cell-free biosynthesis of different high molecular weight forms of bovine neurophysins I and II coded by hypothalamic mRNA. J. Biol. Chem. **254**, 11767–11670 (1979)
5. Drew, D.D.: A mathematical model for prokaryotic protein synthesis. Bull. Math. Biol. **63**, 329–351 (2001)
6. Swain, P.S., Elowitz, M.B., Siggia, E.D.: Intrinsic and extrinsic contributions to stochasticity in gene expression. PNAS **99**, 12795–12800 (2002)
7. Frazier, J.M., Chushak, Y., Foy, B.: Stochastic simulation and analysis of biomolecular reaction networks. BMC Syst. Biol. **3**, 64 (2009)
8. Mier-y-Terán-Romero, L., Silber, M., Hatzimanikatis, V.: The origins of time-delay in template biopolymerization processes. PLoS Comp. Biol. **6**, e1000726 (2010)

9. Fujii, S., Matsuura, T., Sunami, T., Nishikawa, T., Kazuta, Y., Yomo, T.: Liposome display for in vitro selection and evolution of membrane proteins. Nature Protocol **9**, 1578–1591 (2014)

10. Nishimura, K., Matsuura, T., Nishimura, K., Sunami, T., Suzuki, H., Yomo, T.: Cell-free protein synthesis inside giant unilamellar vesicles analyzed by flow cytometry. Langmuir **28**, 8426–8432 (2012)

11. Torre, P., Keating, C.D., Mansy, S.S.: Multiphase water-in-oil emulsion droplets for cell-free transcription? Translation. Langmuir **30**, 5695–5699 (2014)

12. Matsuura, T., Hosoda, K., Kazuta, Y., Ichihashi, N., Suzuki, H., Yomo, T.: Effects of compartment size on the kinetics of intracompartmental multimeric protein synthesis. ACS Synth. Biol. **1**, 431–437 (2012)

13. Murtas, G., Kuruma, Y., Bianchini, P., Diaspro, A., Luisi, P.L.: Protein synthesis in liposomes with a minimal set of enzymes. Biochem. Biophys. Res. Commun. **363**, 12–17 (2007)

14. Kuruma, Y., Stano, P., Ueda, T., Luisi, P.L.: A synthetic biology approach to the construction of membrane proteins in semi-synthetic minimal cells. Biochim. Biophys. Acta **1788**, 567–574 (2009)

15. Sunami, T., Hosoda, K., Suzuki, H., Matsuura, T., Yomo, T.: Cellular compartment model for exploring the effect of the lipidic membrane on the kinetics of encapsulated biochemical reactions. Langmuir **26**, 8544–8547 (2010)

16. Karzbrun, E., Shin, J., Bar-Ziv, R.H., Noireaux, V.: Coarse-grained dynamics of protein synthesis in a cell-free system. Phys. Rev. Lett. **106**, 048104 (2011)

17. Stögbauer, T., Windhager, L., Zimmer, R., Rädlerab, J.O.: Experiment and mathematical modeling of gene expression dynamics in a cell-free system. Integr. Biol. **4**, 494–501 (2012)

18. Mavelli, F., Marangoni, R., Stano, P.: A simple protein synthesis model for the PURE system operation. Bull. Math. Biol. **77**, 1185–1212 (2015)

19. Lazzerini-Ospri, L., Stano, P., Luisi, P.L., Marangoni, R.: Characterization of the emergent properties of a synthetic quasi-cellular system. BMC Bioinfo. **13**, S9 (2012)

20. Calviello, L., Stano, P., Mavelli, F., Luisi, P.L., Marangoni, R.: Quasi-cellular systems: stochastic simulation analysis at nanoscale range. BMC Bioinfo. **14**, S7 (2013)

21. Gillespie, D.T.: Exact stochastic simulation of coupled chemical reactions. J. Phys. Chem. **81**, 2340–2361 (1977)

22. Li, H., Cao, Y., Petzold, L.R., Gillespie, D.T.: Algorithms and software for stochastic simulation of biochemical reacting systems. Biotechnol. Prog. **24**, 56–61 (2008)

23. Cangelosi, D., Fabbiano, S., Felicioli, C., Freschi, L., Marangoni, R.: Quick direct-method controlled (QDC): a simulator of metabolic experiments. EMBnet.journal **19**, 39 (2013)

24. Kubori, T., Shimamoto, N.: Physical interference between Escherichia coli RNA polymerase molecules transcribing in tandem enhances abortive synthesis and misincorporation. N.A.R. **25**, 2640–2647 (1997)

25. Brandt, F., Etchells, S.A., Ortiz, J.O., Elcock, A.H., Hartl, F.U., Baumeister, W.: The native 3D organization of bacterial polysomes. Cell **136**, 261–271 (2009)

26. Luisi, P.L., Allegretti, M., de Souza, T.P., Steiniger, F., Fahr, A., Stano, P.: Spontaneous protein crowding in liposomes: a new vista for the origin of cellular metabolism. Chembiochem **11**, 1989–1992 (2010)

27. De Sousa, P., Stano, P., Luisi, P.L.: The minimal size of liposome-based model cells brings about a remarkably enhanced entrapment and protein synthesis. Chem-BioChem **10**, 1056–1063 (2009)
28. Matsuura, T., Kazuta, Y., Alta, T., Adachi, J., Yomo, T.: Quantifying epistatic interactions among the components constituting the protein translation system. Molec. Syst. Biol. **5**, 297–300 (2011)
29. Nishimura, K., Tsuru, S., Suzuki, H., Yomo, T.: Stochasticity in gene expression in a cell-sized compartment. ACS Synth. Biol. **4**, 566–576 (2015)
30. Murray, C.J., Baliga, R.: Cell-free translation of peptides and proteins: from high throughput screening to clinical production. Curr. Op. Chem. Biol. **17**, 420–426 (2013)
31. Chizzolini, F., Forlin, M., Cecchi, D., Mansy, S.S.: Gene position more strongly influences cell-free protein expression from operons than T7 transcriptional promoter strength. ACS Synth. Biol. **3**, 363–371 (2014)

Self-Organization of a Dichloromethane Droplet on the Surface of a Surfactant Containing Aqueous Solution

Florian Wodlei and Véronique Pimienta[✉]

IMRCP Laboratory, Paul Sabatier University, 118 Route de Narbonne, 31062 Toulouse, France
pimienta@chimie.ups-tlse.fr

Abstract. We investigate the dynamics of a dichloromethane droplet placed on the surface of an aqueous solutions of cetyltrimethylammonium bromide. By varying the surfactant concentration, we observe a rich variety of different shapes, ranging from pulsating over rotating to polygonal-like shaped drops, during the dissolution process. These remarkable shapes seem to be the result of the complex interplay between numerous processes including solubilization, evaporation, mass transfer and adsorption of the surfactants at the water/oil interface.

Keywords: Dichloromethane · Droplets · Cetyltrimethylammonium bromide (CTAB) · Dissolution · Evaporation · Mass transfer · Adsorption of surfactants · Marangoni instability

1 Introduction

Far-from-equilibrium systems show a great variety of spatial and temporal patterns, which are known as dissipative structures. The interplay between physico-chemical processes and mass or heat transfer can give rise to convective flows that may form such structures (Davies and Rideal 1963). The origin of dissipative structures, such as regular convective cells, interfacial deformations or interfacial turbulences, is often the conversion from chemical to mechanical energy (Eckert et al. 2012).

Specially the flows, which are triggered by density or surface tension gradients, are crucial to many natural phenomena, like ocean and atmospheric flows (Budroni et al. 2012). They are also crucial to processes like extraction, spreading of spills in aquifers or oil recovery and of course crucial to chemical reactors at all scales.

The energy transduction, that is created when the size of the phases in contact is very different, can give rise to spontaneous motion of the smaller phase (de Gennes et al. 2003). This is the case for liquid drops on solid or liquid surfaces (Zhao and Pumera 2012; Pimienta and Antoine 2014).

Symmetry breaking, which is inherent of motion, is obtained on solid surfaces by external constraints like thermal gradients or concentration gradients

© Springer International Publishing Switzerland 2016
F. Rossi et al. (Eds.): WIVACE 2015, CCIS 587, pp. 159–170, 2016.
DOI: 10.1007/978-3-319-32695-5_15

of imprinted chemicals on the surface (Chaudhury and Whitesides 1992). On liquid surfaces motion can be induced spontaneously by the drop itself. Such autonomous behaviour is at the origin of the interest for such systems in relation with artificial cell design. There, by observing and mimicking the shapes and trajectories, which are spontaneously chosen in biological systems to adapt to motility (Bush and Hu 2006), the aim is to identify the energy sources that drive the different propulsion mechanisms. Motion is one of the vital functions of microorganisms in search of new resources.

The system of interest here is a drop of dichloromethane (DCM) deposited on a cetyltrimethylammonium bromide (CTAB) solution. By changing the surfactant concentration, which plays the role of a control parameter, the drop shows a succession of well defined, highly ordered patterns (Pimienta et al. 2011).

In the following, we will first go back to earlier observations, obtained under biphasic conditions, which have caused our interest for this system. In the second part of the article, we will describe the main regimes observed in the drop geometry.

2   Experimental Part

2.1   Biphasic System

Cetyltrimethylammonium bromide (CTAB) (Aldrich, $\geq 99\%$), dichloromethane (DCM, $\rho = 1.33\,\mathrm{g \cdot cm^{-3}}$) (Aldrich, HPLC grade) aswell as ultra-pure water (resistivity $\geq 18\,\mathrm{M\Omega \cdot cm}$) for preparing the solutions were used. All chemical reagents used were of analytical grade.

Oscillations were recorded in a glass beaker with a 31 mm inner diameter without stirring. 15 mL of the organic solution (DCM) was placed at the bottom of the beaker. Then, 15 mL of aqueous solution containing CTAB was carefully placed on top of the organic phase.

Surface tension measurements were performed using a small cylinder (diameter 2.8 mm; 10 mm high) made of high-density polyethylene and connected to a microbalance (Fig. 1).

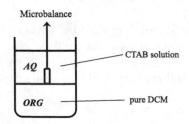

Fig. 1. Sketch of the setup for the system in the biphasic configuration in a 30 mL beaker.

2.2   Drop Configuration

The same reactants mentioned above were used for this configuration. 25 mL of the aqueous solution containing CTAB were poured in a Petri dish of 7 cm diameter. Then, a drop of 25 μL of the organic solution (DCM) was placed carefully on the surface of the aqueous solution with a HPLC micro-syringe. After addition, the Petri dish was covered.

2.3   Visualization Techniques in the Drop Configuration

To visualize the droplet on the solution two techniques were used. The first, called *Shadowgraphy*, is realized by the illumination of the system from the top with a collimated light beam (a LED light source with fibre connection was used). Then, the projected shadow, that is produced by the highly curved regions at the corner of the droplet, is recorded. The images, which are shown in the overview of the different shape regimes (Fig. 11), were recorded in this way.

The other technique used, called *Schlieren technique*, allows a better contrast of the drop. To obtain this, a sharp edge was placed between a LED back-light illumination and a focusing lens, exactly at the focal point of the lens. This creates an intensity gradient on the recorded image. This density gradient can be varied by changing the distances and the form of the sharp edge (Fig. 2). In all other images shown in this work this techniques has been used.

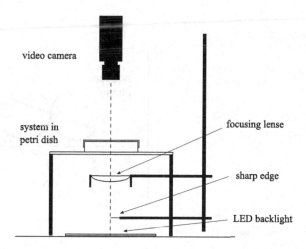

Fig. 2. Sketch of the optical setup for the Schlieren technique with the system in the drop configuration in the Petri dish.

3   Biphasic System

The first theoretical study aimed to establish criteria predicting Marangoni instability onset during solute mass transfer in biphasic systems was performed by

Sternling and Scriven (1959). They applied linear stability analysis to systems where a solute is transferred through a non-deformable interface between two semi-infinite liquid layers. The instability can develop in systems far from partition equilibrium and its appearance depends mainly on the solvent properties, the surface activity of the solute and on the formation of critical solute concentration gradients in the normal to the interface direction (Kovalchuk et al. 2012). Unstable transfer is expected when the solute diffuses out of the phase in which its diffusivity is lower and kinematic viscosity is higher.

The biphasic configuration of the present system, where, as described in the experimental part, an aqueous solution of CTAB is placed on the top of the heavier pure DCM phase fulfills the above criteria. The measured interfacial tension between the organic and the aqueous phase showed relaxation oscillations shown in Fig. 3 (Lavabre et al. 2005).

It was shown that this oscillations are induced by a periodic Marangoni instability. The decrease of the surface tension started just when movements at the interface were initiated (see vertical line in inset of Fig. 3).

The oscillations observed arise as a result of the mass transfer of CTAB (Fig. 3, B). Under mass transfer, concentration gradients build up in the diffusion layer (Pradines et al. 2007). The surfactant concentration in contact with the interface (that determines the corresponding adsorption level) decreases.

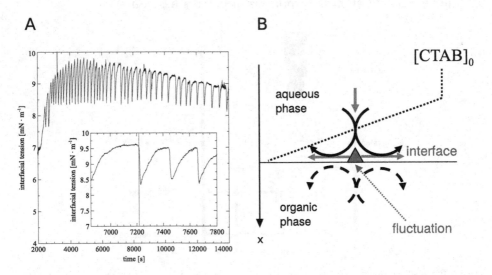

Fig. 3. A: oscillations of the interfacial tension between DCM and an aqueous solution containing 5 mM CTAB. Inset shows the oscillations between 6850–7800 sec in detail. The line in the inset indicates the time when movements at the interface were initiated. B: Schematic representation of the processes leading to oscillations in the biphasic system (details see text). Dashed line represents the normal to the interface concentration gradient of the surfactant, green triangle stands for the initial fluctuation, red arrows represent Marangoni flows and the curved arrows represent the convective cells induced by the Marangoni instability (Colour figure online)

At the interface, concentration heterogeneities are likely to occur, leading to tangential concentration gradients that induce stretching of the interface due to the Marangoni effect. This radial flow creates in turn a vertical flow that brings aqueous solution from the bulk, richer in surfactants, to the interface and will amplify the initial surface heterogeneities to give rise to convective flows in both phases. The related intense mixing of the aqueous layer near the interface destroys the normal gradients, the instability vanishes, and the system switches to a slower, diffusive process. During this phase, normal gradients, induced by CTAB mass transfer, built up again and the cycle starts anew.

4   Drop Configuration

In this configuration, where, as described in the experimental part, a drop of DCM is placed on the surface of an aqueous phase containing CTAB, there are other processes in addition to CTAB mass-transfer. Due to the low boiling point ($T_{bp} = 39.6\,°C$ at 1 atm) and the partial miscibility of DCM (solubility in water is 0.15 M), evaporation and dissolution of DCM in the aqueous phase are also at play in this geometry.

The evolution of the drop shows two phases, an induction period, during which the drop shrinks while maintaining its circular shape, and a second phase at which the drop starts to deform in a self-organized manner before its final disappearance.

4.1   Induction Period

We measured the initial diameters of the deposited drop as a function of the CTAB concentration and observed the same trend as for the air/water interfacial tension (Fig. 4). Below the critical micellar concentration (cmc = 0.9 mM), $\gamma_{w/a}$ increases as does the initial diameter. After the cmc, both $\gamma_{w/a}$ and the initial diameter remain constant.

If we assume a spherical cap geometry for the heavy DCM drop, the forces acting on the triple contact line can be represented by the Neumann triangle as shown in Fig. 5. The corresponding young equation reads $\gamma_{w/a} = \gamma_{o/a} + \gamma_{w/o} \cdot \cos\theta$.

A decrease of the water/air interfacial tension, $\gamma_{w/a}$, results in this case in a more compact drop (decreased diameter and increased contact angle), which is what we observe experimentally.

We have also measured the time span of the induction period for a 25 μL DCM drop. For 0.25 mM and 0.5 mM, no induction time is observed, the unstable regime starts immediately. The longest induction period is observed at 1 mM, while its duration decreases again with increasing CTAB concentration.

In correlation to the variation of the initial diameter it appears that flatter drops (i.e. before the cmc) immediately enter the unstable regime while thicker drops need to reach a critical size before the instability is triggered. It looks as if the drop has to be small (flat) enough for the instability to start. In other words, surface tension forces need to overcome gravity effects in order to trigger the Marangoni instability.

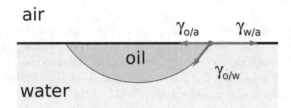

Fig. 4. Initial diameter (A), water/air interfacial tension (B) and induction time (C) as a function of the CTAB concentration.

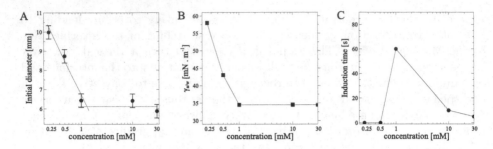

Fig. 5. Forces acting on a liquid droplet on a liquid surface in a spherical cap approximation according to the Young equation.

The decrease of the induction time for higher CTAB concentrations may be related to an increased solubilization rate due to the formation of CTAB swollen micelles (i.e. the formation of an oil in water microemulsion) (Tadmouri et al. 2008).

4.2   Pattern Formation - Marangoni Instability

Spreading and Translation. Examples of motions of drops on liquid surfaces are known in literature like the pentanol droplet of Nagai et al. (2005) or the aniline droplet of Chen et al. (2009), both on aqueous solutions. The aniline drop is similar to the present system in the sense that dissolution, evaporation and density effects are also involved. In this case evaporation and solubilization are relatively slow and the drop appears as a macroscopic drop coexisting with a surrounding precursor film. Two different kinds of motion are observed: bee-like motion and circular motion that could last for hours. Motility was ascribed to a surface tension imbalance at the front and rear of the drop. This difference, revealed by a difference in contact angle, was attributed to the precursor film. The film is driven by a Marangoni flow and may lose its initial symmetry (by fluctuations or induced manually). The initiated translation is then sustained by the dissymmetry of the surface active film pointing to the rear of the drop.

For the DMC drop on the CTAB solution, the film and its distortion are eye visible (Fig. 6). Motion is however observed only transiently, for 3 or 4 s (the whole process lasts for about 20 s). The main regime appears as a drop

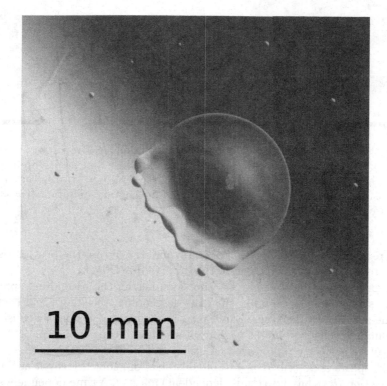

Fig. 6. Translation of a DCM drop on a 0.25 mM CTAB solution.

surrounded by a thin film terminated by a ring of smaller droplets. At some point, the film recoils on one side of the drop pushing the drop on the opposite side. Motion gets started by the induced dissymmetry. While the asymmetric shape is maintained, circular or translational motion are observed.

Pulsations. To our knowledge, there are two examples of similar pulsations in the literature. Bates et al. (2008) report on a pulsating regime, of a very short life time, that gives rise to the ejection of a ring of droplets for a 1-butanol drop cooled to 1 °C deposited on a 40 °C water phase. Another example is a mineral oil drop containing a non-ionic, non-water soluble surfactant placed on water, which is reported by Stocker and Bush (2007). The time scale here is longer (tens of seconds) but the recorded pulsations are very irregular compared to the one in the present system.

For a CTAB concentration of 0.5 mM, pulsations can last between 5 and 20 s with a mean frequency of about 1 Hz. Each pulsation consists of a rapid spreading of the drop. At maximum expansion, which represents close to 25 % of the initial drop radius, the rim formed at the edge of the expanding film undergoes a Rayleigh-Plateau-like instability and breaks up into small droplets that move radially away from the drop (over a distance of 5 to 6 mm) and rapidly disappear by evaporation and dissolution. After detachment of the rim the remaining film recedes to form a compact drop again (Fig. 7).

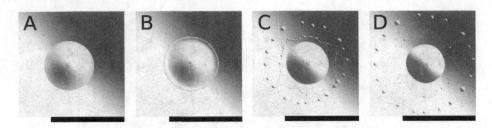

Fig. 7. Pulsating DCM drop on a 0.5 mM CTAB solution. A: Drop starts to expand (t = 0 s). B: Rim is forming and expanding (t = 160 ms). C: Rim breaks up in several droplets while the film between rim and droplet is receding (t = 345 ms). D: Drop has ejected his rim completely and is about to start for another pulsation (t = 435 ms). Black bar corresponds to 10 mm.

The regularity and conservation of the symmetry during the beating pattern are reflected in the time evolution of the drop radius (Fig. 8).

The recorded oscillations are not time-symmetric: the ascending, spreading parts are slightly longer (~ 0.6 s) than the descending, receding parts (~ 0.4 s). This time-asymmetry increases significantly in the course of time. By considering a very simple mechanical model of rim pulsations the time evolution of the drop radius can be perfectly reproduced (Antoine and Pimienta 2013).

The drop rim is described as a toroidal tube of oil, that is located at the drop radius position $R(t)$, having a (time-dependent) mass m. A time-dependent spring parameter $k(t)$, that takes all the springlike forces acting on the rim into account, is used. These forces are surface tensions, Marangoni stress, as well as the drop surface elasticity due to the surfactant monolayer. The spring parameter can be seen as the opposite of an effective spreading coefficient ($S_{\text{eff}}(t) = -k(t)/2\pi$).

The model also takes into account the hydrostatic force acting on the rim and also a kind of damping force, that is caused by the friction of the toroidal rim. Finally, by summing up all forces, the dynamics of the rim is described by:

$$\frac{d}{dt}\left(m(t)\frac{dR}{dt}\right) = k(t)R + k_0\frac{R_0^4}{R^3} - bR\frac{dR}{dt}$$

A constant mass m_0 and a decreasing spring parameter $k(t)$ is considered for the mean ascending part of $R(t)$ and a constant spring parameter k_0 and an increasing mass $m(t)$ for the mean receding part. The whole curve of $R(t)$ can be simulated by choosing a simple rim breakup criterion given by the capillary length. To account for the slight period increase observed, a slight increase of the rim mass over all of the pulsations is included.

In other words, the above scenario can be interpreted as follows: the oscillations are the result of Marangoni induced spreading due to a non homogeneous distribution of the CTAB at the water/oil interface. During expansion gradients are enhanced and spreading is amplified. This leads to a critical thickness of the film at the rear of the rim, which detaches while the film dewets to reform the initial drop.

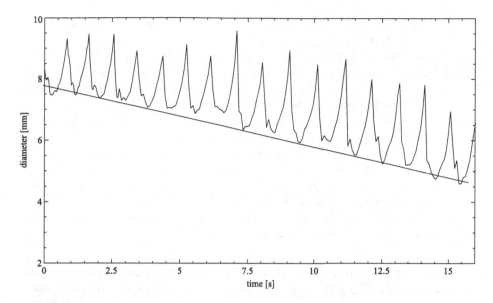

Fig. 8. Regular oscillations of the diameter of a pulsating DCM droplet on a 0.5 mM CTAB solution. The global linear decrease of the drop diameter can be attributed to the mass loss of the droplet.

Rotation. For a CTAB concentration of 10 mM, the drop undergoes a major shape deformation after the induction period, leading to an elongated structure with two sharp tips (Fig. 9). This structure then rotates in an arbitrary (clockwise and counterclock-wise rotations are equally observed) but constant direction. The rotating two-armed drop is the most frequently encountered structure, but also rotating drops with three or sometimes four tips are observed.

Typically a drop can perform between 30 and 60 rotations with a steady rotation frequency of around 1.9 Hz, which is independent of the drop volume and time. During this stable rotation, the drop ejects smaller droplets from its tips. Each daughter droplet moves approximately 4 mm, flattens slightly, and then decays into several smaller droplets, that finally disappear.

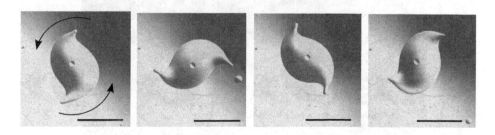

Fig. 9. Rotating of a DCM drop on a 10 mM CTAB solution (time increases from left to right). Approx. 1 rotation is shown (with 200 ms between each image). Black bar corresponds to 5 mm.

Although there are no examples of rotating drops in literature, there are rotating solids and gels on liquid phases (e.g. Nakata et al. 1997; Sharma et al. 2012).

In the article by Nakata et al. a solid camphor fragment with a comma-shaped form is rotating. The authors explain the rotation in the same way as the pure translation of the aniline drop, which is discussed above. The dissolving camphor forms a thin layer at the interface which locally reduces the interfacial tension. In the concave region of the "rotor", due to the curvature, there is an accumulation of dissolved camphor in contrast to the convex region. Therefore a gradient in the surface tension is created and the rotor moves with the convex region ahead and the fragment starts to rotate.

For the DCM drop the same mechanism might be at play. Moreover the constant rotation frequency could be explained by the fact that the concentration in the concave regions of the rotors of the drop are continuously renewed from the bulk solution of the droplet, which is pure DCM.

Polygonal-Like Shape. For higher CTAB concentrations, a polygonal-like shape is observed (Fig. 10). After the induction period small tips form along the edge of the drop, which confers the drop the shape of a polygon. This tips move and when two of them collide they can give rise to the ejection of a smaller

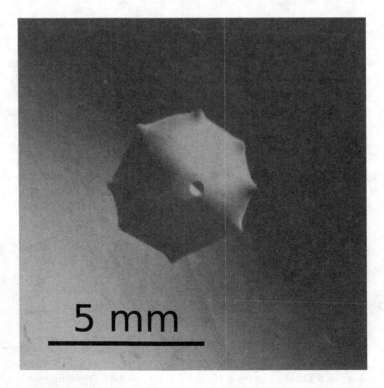

Fig. 10. Polygonal-like shaped drop on a 30 mM CTAB solution.

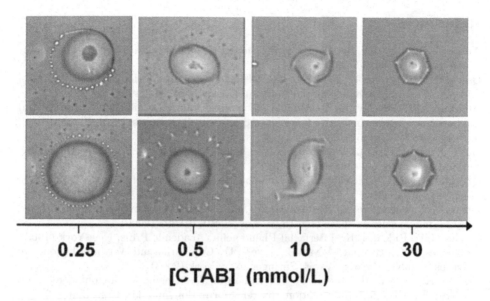

Fig. 11. Overview of the different regimes as a function of the CTAB concentration and time (vertical axis shows the time evolution).

droplet. During shrinkage of the drop the number of tips is reduced until the final disappearance of the drop. At the moment, we have no insight on this behaviour, which, to our knowledge, has never been described in the literature.

5   Conclusion and Perspectives

We described here the complex shape evolution of a DCM droplet on the surface of a CTAB containing aqueous solution. An overview of the different shape regimes is given in Fig. 11 as a function of the CTAB concentration. The phenomena observed range from spreading and dewetting below the cmc to rotating and polygonal-like shaped drops above the cmc.

The system described here is chemically extremely simple and easy to handle but gives rise to complex coupled physical processes. The surfactant concentration, used here as a control parameter, induces an amazing range of shapes and motion patterns. Coupled to these shape-forming processes is the emission of very small but macroscopic droplets. This system is the first example of such a sequence of highly ordered patterns induced by coupled hydrodynamic instabilities. The resulting structures show very efficient motility, internal agitation and dispersion properties.

We believe that this system for which the transitions between several regimes are driven by a controlled and progressive change of the physico-chemical properties, offers a great opportunity of a step forward in the understanding and modeling of fundamental knowledge in the very broad field of convective instabilities.

References

Antoine, C., Pimienta, V.: Mass-spring model of a self-pulsating drop. Langmuir **29**(48), 14935–14946 (2013)

Bates, C.M., Stevens, F., Langford, S.C., Dickinson, J.T.: Motion and dissolution of drops of sparingly soluble alcohols on water. Langmuir **24**, 7193–7199 (2008)

Budroni, M.A., Rongy, L., De Wit, A.: Dynamics due to combined buoyancy- and Marangoni-driven convective flows around autocatalytic fronts. Phys. Chem. Chem. Phys. **14**, 14619–14629 (2012)

Bush, J.W.M., Hu, D.L.: Walking on water: biolocomotion at the interface. Annu. Rev. Fluid Mech. **38**, 339–369 (2006)

Chaudhury, M.K., Whitesides, G.M.: How to make water run uphill. Science **256**(5063), 1539–1541 (1992)

Chen, Y.-J., Nagamine, Y., Yoshikawa, K.: Self-propelled motion of a droplet induced by marangoni-driven spreading. Phys. Rev. E **80**, 016303 (2009)

Davies, J.T., Rideal, E.K.: Interfacial Phenomena. Academic Press, New York (1963)

de Gennes, P.-G., Brochard-Wyart, F., Quere, D.: Capillarity and Wetting Phenomena: Drops, Bubbles, Pearls, Waves. Springer, New York (2003)

Eckert, K., Acker, M., Tadmouri, R., Pimienta, V.: Chemo-Marangoni convection driven by an interfacial reaction: pattern formation and kinetics. Chaos **22**(3), 037112 (2012)

Kovalchuk, N.M., Pimienta, V., Tadmouri, R., Miller, R., Vollhardt, D.: Ionic strength and pH as control parameters for spontaneous surface oscillations. Langmuir **28**(17), 6893–6901 (2012)

Lavabre, D., Pradines, V., Micheau, J.C., Pimienta, V.: Periodic Marangoni instability in surfactant (CTAB) liquid/liquid mass transfer. J. Phys. Chem. B **109**(15), 7582–7586 (2005)

Nagai, K., Sumino, Y., Kitahata, H., Yoshikawa, K.: Mode selection in the spontaneous motion of an alcohol droplet. Phys. Rev. E Stat. Nonlin. Soft Matter Phys. **71**(6 Pt 2), 065301 (2005)

Nakata, S., Iguchi, Y., Ose, S., Kuboyama, M., Ishii, T., Yoshikawa, K.: Self-rotation of a camphor scraping on water: new insight into the old problem. Langmuir **13**(16), 4454–4458 (1997)

Pimienta, V., Antoine, C.: Self-propulsion on liquid surfaces. Curr. Opin. Colloid Interface Sci. **19**, 290–299 (2014)

Pimienta, V., Brost, M., Kovalchuk, N., Bresch, S., Steinbock, O.: Complex shapes and dynamics of dissolving drops of dichloromethane. Angew. Chem. Int. Ed. **50**, 10728–10731 (2011)

Pradines, V., Tadmouri, R., Lavabre, D., Micheau, J.C., Pimienta, V.: Association, partition, and surface activity in biphasic systems displaying relaxation oscillations. Langmuir **23**, 11664–11672 (2007)

Sharma, R., Chang, S.T., Velev, O.D.: Gel-based self-propelling particles get programmed to dance. Langmuir **28**(26), 10128–10135 (2012)

Sternling, C.V., Scriven, L.E.: Interfacial turbulence: hydrodynamic instability and the marangoni effect. AIChE J. **5**, 514–523 (1959)

Stocker, R., Bush, J.W.M.: Spontaneous oscillations of a sessile lens. J. Fluid Mech. **583**, 465–475 (2007)

Tadmouri, R., Zedde, C., Routaboul, C., Micheau, J.C., Pimienta, V.: Partition and water/oil adsorption of some surfactants. J. Phys. Chem. B **112**(39), 12318–12325 (2008)

Zhao, G., Pumera, M.: Macroscopic self-propelled objects. Chem. Asian J. **7**, 1994–2002 (2012)

From Microscopic Compartmentalization to Hydrodynamic Patterns: New Pathways for Information Transport

Marcello A. Budroni[1]([✉]), Jorge Carballido-Landeira[2], Adriano Intiso[3],
Lorena Lemaigre[2], Anne De Wit[2], and Federico Rossi[3]([✉])

[1] Department of Chemistry and Pharmacy, University of Sassari, Sassari, Italy
mabudroni@uniss.it
[2] Nonlinear Physical Chemistry Unit, Service de Chimie Physique et Biologie
Théorique, Université libre de Bruxelles, CP 231 - Campus Plaine,
1050 Brussels, Belgium
[3] Department of Chemistry and Biology, University of Salerno,
via Giovanni Paolo II 132, 84084 Fisciano, SA, Italy
frossi@unisa.it
http://physchem.uniss.it/cnl_dyn/budroni.html,
http://www.unisa.it/doconti/federicorossi/index

Abstract. Can we exploit hydrodynamic instabilities to trigger an efficient, selective and spontaneous flow of encapsulated chemical information? One possible answer to this question is presented in this paper where cross-diffusion, which commonly characterizes compartmentalized dispersed systems, is shown to initiate buoyancy-driven hydrodynamic instabilities. A general theoretical framework allows us to predict and classify cross-diffusion-induced convection in two-layer stratifications under the action of the gravitational field. The related nonlinear dynamics is described by a cross-diffusion-convection (CDC) model where fickian diffusion is coupled to the Stokes equations. We identify two types of hydrodynamic modes (the negative cross-diffusion-driven convection, NCC, and the positive cross-diffusion-driven convection, PCC) corresponding to the sign of the cross-diffusion term dominating the system dynamics. We finally show how AOT water-in-oil reverse microemulsions are an ideal model system to confirm the general theory and to approach experimentally cross-diffusion-induced hydrodynamic scenarios.

Keywords: Transport of chemical information · Compartmentalization · Water-in-oil reverse microemulsions · Buoyancy-driven instabilities · Cross-diffusion · Multi-components systems

1 Introduction

The compartmentalization of chemical systems plays a central role in generating unexpected novelties and emergent dynamical behaviors. Many self-assembled matrices including micelles [11,12], lipid vesicles [20] and microemulsions [25]

© Springer International Publishing Switzerland 2016
F. Rossi et al. (Eds.): WIVACE 2015, CCIS 587, pp. 171–183, 2016.
DOI: 10.1007/978-3-319-32695-5_16

have been extensively studied as model systems for mimicking biological cellular environments and related complex intra- and inter-dynamics. The spatial confinement of chemical species can affect the global dynamics of a system by changing drastically the diffusive transport inside and across the confined domains and also inducing a catalytic effect in the chemical reactivity [22]. For these characteristics, dispersed media have often been studied in combination with nonlinear oscillatory kinetics to approach collective behaviors, communication and synchronization dynamics in networks of coupled inorganic oscillators [6,13,16,19,21].

In this context, the AOT (sodium bis(2-ethylhexyl)sulfosuccinate Aerosol OT) water-in-oil reverse microemulsions (ME) are one of the most investigated systems. ME are liquid mixtures of an organic solvent (more often termed oil) in which water domains surrounded by a surfactant (AOT) can form a thermodynamically stable dispersion (see Fig. 1). In particular, under the percolation threshold, ME appear at the nano-scale as dispersed spherical or elongated droplets where the surfactant constitutes a sort of membrane with the hydrophobic part oriented towards the outer organic phase and the hydrophilic heads in contact with the inner aqueous phase segregated into the droplet. In this situation, the water core of the ME can be conveniently used to dissolve hydrophilic molecules or reactants and become a suitable tool for nano-synthesis, drug delivery or for studying chemical communication [6,8]. The oscillatory Belousov-Zhabotinsky (BZ) reaction has been thoroughly studied in ME. Here BZ reagents are solubilized in the water core of AOT-coated nano-droplets to create a great amount of nano-oscillators, which dynamics is coupled *via* some active intermediates of the BZ system able to cross the hydrophobic membrane of the surfactant and act as messengers among the confined water domains. Microemulsions are also important in the realm of reaction-diffusion-driven pattern formation. Turing structures and other exotic dissipative structures have been found when ME, loaded with reactants of the BZ system, are studied in spatially extended reactors [7,24]. Thanks to the selective permeability of the surfactant barrier with regard to the BZ species, this system sustains differential diffusion and can meet the requirements for a Turing instability with the slow diffusion mode involving the activator of the BZ oscillator (which moves with the droplets), while the hydrophobic inhibitor, free to move alone through both the oil and the aqueous phase, presents a diffusion coefficient two orders of magnitude larger than the activator.

Cross-diffusion, whereby a flux of a given species entrains the diffusive transport of another species, is also an important process which takes place in ME systems. Measurements of cross-diffusion coefficients in ternary AOT microemulsions (H_2O (1)/AOT (2)/oil) revealed that the cross-diffusion coefficient D_{12}, which describes the flux of water induced by a gradient in the surfactant concentration, can be significantly larger than both D_{11} and D_{22}, i.e. the main diffusion coefficients of water and AOT, respectively [9,10]. The crucial influence of cross-diffusive contributions in terms of pattern formation has been proved in non-reactive and reactive spatially distributed systems, both theoretically [26,27] and experimentally [4,14]. Thus, the constraints imposed to diffusive transport

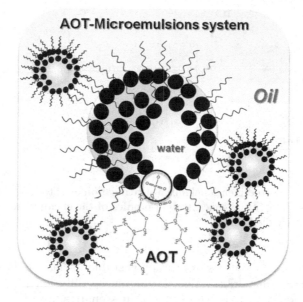

Fig. 1. Sketch of AOT water-in-oil reverse microemulsions.

by compartmentalization at a microscopic scale impact emergent behaviors at a macroscopic characteristic length. Apart for reaction-diffusion patterns, such variety of diffusive modes can trigger hydrodynamic instabilities, when buoyancy forces are at play [23]. Convection is the most efficient spontaneous transport mechanism and can be suitably engineered to enhance spatial spreading of chemical information encapsulated into droplets in response to initial concentration gradients.

With this perspective, we review our recent work where ME are presented as a convenient model system for inducing cross-diffusion-driven hydrodynamic flows in double-layer systems. We first present a general modeling and theory on buoyancy-driven convection promoted by cross-diffusion and we successively study experimentally convective patterns occurring when two ME are stratified in a Hele-Shaw cell [2,3].

2   Theory

Consider a two dimensional vertical slab of width L_X and height L_Z in a (X, Z) reference frame, where the gravitational acceleration $\bar{g} = (0, -g)$ is oriented downwards along the Z axis. The solution T of density ρ^T, containing the solute h with the initial concentration $C_{h,0}^T$ and the solute j with concentration $C_{j,0}^T$ is placed on top of the solution B, with concentration $C_{h,0}^B = C_{h,0}^T$, $C_{j,0}^B > C_{j,0}^T$ and density $\rho^B > \rho^T$ (see sketch in Fig. 2). In other words, the species h is homogeneously distributed over the spatial domain, while the concentration of the species j increases downwards the gravitational axis.

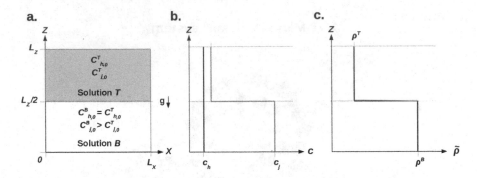

Fig. 2. Sketch of the two-dimensional initial stratification used to study cross-diffusion-driven convection. The initial concentration profiles of the chemical solutes result in a step function density distribution.

The resulting double-layer stratification is stable to classical Rayleigh-Taylor or buoyancy-driven instabilities due to differential diffusion mechanisms, such as double-diffusion or double-layer-convection scenarios, and thus it is ideal to isolate the pure effect of cross-diffusion on the system stability.

Before the onset of an instability, we can assume that the flow is at rest and that the concentration profiles of the species do not vary along the horizontal x direction. The initial evolution of the concentration fields can be thus followed along the vertical coordinate z and described by means of the dimensionless cross-diffusion equations

$$\partial_\tau c_j = \nabla^2 c_j + \delta_{jh} \nabla^2 c_h \tag{1}$$
$$\partial_\tau c_h = \delta_{hj} \nabla^2 c_j + \delta_{hh} \nabla^2 c_h \tag{2}$$

where the spatial and the time variables are scaled by L_0 and $t_0 = L_0^2/D_{jj}$, respectively and D_{jj} is the dimensional main diffusivity of the heterogeneously distributed species. The concentration fields are non-dimensionalized by the reference $\Delta C_{j,0} = (C_{j,0}^B - C_{j,0}^T)$ according to $(c_j, c_h) = (C_j - C_{j,0}^T, C_h - C_{h,0}^T)/\Delta C_{j,0}$, and $\delta_{hh} = D_{hh}/D_{jj}$ is the ratio between the main molecular diffusion coefficient of solute h to that of j. Similarly, $(\delta_{jh}, \delta_{hj}) = (D_{jh}, D_{hj})/D_{jj}$.

Possible cross-diffusion feedbacks on the dynamics are measured in the matrix δ and, in particular, due to the sharp initial gradient imposed to the concentration profile $c_j(z,0)$, the off-diagonal term δ_{hj} dominates the initial part of the dynamics while the other cross-diffusivity, δ_{jh}, plays a negligible role. If δ_{hj} is positive, the species j, free to diffuse from the bottom to the upper layer in response to its initial concentration gradient (Fig. 3(a)), generates a co-flux in h and, as a result, the initially flat concentration profile $c_h(z,\tau)$ develops a non-monotonic shape with a local maximum and minimum symmetrically located above and below the initial interface, respectively (Fig. 3(b)). By contrast, the propagation of solute j towards the upper layer triggers a counter-flux in h if δ_{hj} is negative. In the concentration profile $c_h(z,\tau)$ this produces in time a local depletion area in the upper layer and an accumulation just below the initial interface located at $z_0 = L_Z/(2L_0)$ (Fig. 3(d–e)). We clearly observe an inversion in the morphology of the non-monotonic profiles when switching from a positive to a negative δ_{hj} and the relative intensity of the extrema along the non-monotonic profiles reflects the magnitude of this term.

The non-monotonic concentration profiles triggered by cross-diffusion in species h can drastically impact the density distribution along the gravitational axis. The density profile can be computed as a Taylor expansion of the concentration fields according to

$$\rho(z, \tau) = c_j(z, \tau) + R\, c_h(z, \tau) \tag{3}$$

where the buoyancy ratio

$$R = \frac{\alpha_h}{\alpha_j} \tag{4}$$

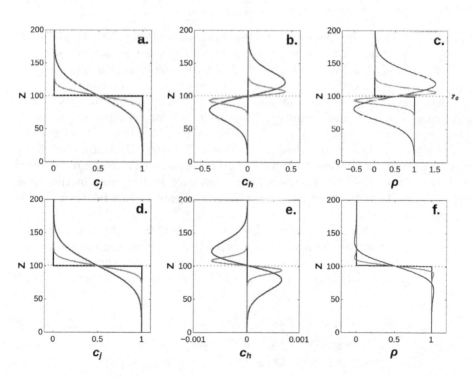

Fig. 3. Typical spatio-temporal evolution of the dimensionless concentration profiles $c_j(z, \tau)$ and $c_h(z, \tau)$ when $\delta_{hj} > 0$ (4.70) (panels (a) and (b)) or $\delta_{hj} < 0$ (−0.01) (panels (d) and (e)); panels (c) and (f) show the corresponding dimensionless density profiles computed by means of Eq. 3, with $R = 0.5$ and 100, respectively. In each panel black lines describe the initial distribution of the species, while red and blue profiles depict the spatial concentrations at successive times (Colour figure online).

quantifies the relative weight of the initially homogeneous species, h, to the global density with respect to species j, featuring the initial concentration jump. In Eq. (3), $\alpha_j = \frac{1}{\rho^T} \frac{\partial \tilde{\rho}}{\partial C_j}$ and $\alpha_h = \frac{1}{\rho^T} \frac{\partial \tilde{\rho}}{\partial C_h}$ the solutal contributions of the species j and h, respectively, to the dimensional density $\tilde{\rho}$.

When the contribution to the density of the initially homogeneous species is dominant, the shape of $c_h(z, \tau)$, will be reflected in the spatio-temporal evolution of the density profile $\rho(z, \tau)$. Under the action of the gravitational field, non-monotonic density profiles are at the basis of convective instabilities (see Fig. 3(c) and (f)), whereby denser areas of a fluid locally overlies less dense zones.

By analyzing the morphology of a density profile along the gravitational axis we can, thus, predict conditions for the onset of cross-diffusion-driven hydrodynamic instabilities together with the topology of the resulting patterns [2,3,23]. In recent work [1], it has been demonstrated that conditions for a convective instability are met when

$$R > \frac{\sqrt{\delta_{hh}}(1 + \sqrt{\delta_{hh}})}{|\delta_{hj}|} \tag{5}$$

and the sign of δ_{hj} discriminates the domains where Positive Cross-Diffusion-Driven Convection, **PCC** scenarios ($\delta_{hj} > 1$) or Negative Cross-diffusion-driven Convection, **NCC** scenarios ($\delta_{hj} < 1$) are to be expected. In particular, in PCC scenarios, starting from an initially stable condition in which the density increases downwards the gravitational axis, the density profile changes in time developing a density maximum overlying a density depletion area across the initial interface between the two layers. This results in a fingered deformation of the interface, like in the classical double diffusion (DD) instability [5,23]. NCC profile is reminiscent of typical density profiles characterizing diffusive layer convection (DLC) scenarios where convective modes grow in the upper and the lower layer, without deforming the initial contact line between the two stratified solutions [5,23].

In order to confirm the theoretical results obtained from the density-profile-based analysis and to obtain a spatio-temporal picture on the cross-diffusion-driven convection phenomenology, it is necessary to solve the full nonlinear problem in which Fickian equations are coupled to Stokes equations (for details see [1]):

$$\partial_\tau c_j + (\mathbf{v} \cdot \nabla) c_j = \nabla^2 c_j + \delta_{jh} \nabla^2 c_h \tag{6}$$

$$\partial_\tau c_h + (\mathbf{v} \cdot \nabla) c_h = \delta_{hj} \nabla^2 c_j + \delta_{hh} \nabla^2 c_h \tag{7}$$

$$\nabla p = \nabla^2 \mathbf{v} - (R c_h + c_j) \mathbf{1}_z \tag{8}$$

$$\nabla \cdot \mathbf{v} = 0. \tag{9}$$

The CDC equations are derived in the dimensionless form by introducing the set of scaled variables $\{\tau = t/t_0, (x, z) = (X, Z)/L_0, (c_j, c_h) = (C_j - C_{j,0}^T, C_h - C_{h,0}^T)/\Delta C_{j,0}, \mathbf{v} = \mathbf{V}/v_0, p = P/p_0\}$, where the scales of the velocity $\mathbf{v}$ and the pressure p are defined as $v_0 = L_0/t_0$ and $p_0 = \mu/t_0$, respectively.

By solving Eqs. (6–9) through the Alternating Direction Implicit Method (ADI), we can obtain an overview on the dynamics of the two possible instabilities. Simulations are run by using no-flux boundary conditions for the concentration field of the chemical species at the four solid boundaries of the squared spatial domain while no-slip boundary conditions are imposed for the velocity field. Figure 4 shows the dynamical destabilization of the two-layer interface due to a PCC mechanism (panel (a)) and a typical NCC scenario (panel (b)). The snapshots illustrate the spatio-temporal evolution of the instabilities by mapping the vorticity $\omega(x, z, \tau) = \nabla \times \mathbf{v}$ over the simulation spatial domain. In both cases, the unstable area starts from the border of the spatial domain where a numerical perturbation can break the symmetry and extends along the horizontal direction. As convective fingers form, they grow vertically along the gravitational axis. Delayed forming fingers show an apparent drifting towards the side where the instability nucleates, attracted by residual flows and progressively merge with pre-existing fingers.

The PCC scenario is induced by the positive cross-diffusion term δ_{hj} when the buoyancy ratio R meets the requirement of Eq. (5). As previously shown in Fig. 3(a) and (b), solute j diffuses form the bottom to the upper layer due to the initial gradient and triggers a non-monotonic distribution of $c_h(z, \tau)$, thus inducing a local density maximum over a minimum downwards z, symmetrically located around z_0 (Fig. 3(c)).

Again, when Eq. (5) is satisfied, a negative cross-diffusion coefficient δ_{hj} initiates a NCC-type convective pattern. Here the motion of solute j sustains the non-monotonic concentration profile $c_h(z, \tau)$ shown in Fig. 3(e). During the development of the instability the initial interface is not deformed because of the formation of the density maximum located below the initial interface, which acts as a density barrier preventing the finger growth from the top to the bottom layer (Fig. 3(f)).

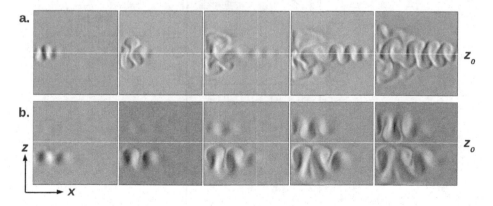

Fig. 4. Typical spatio-temporal evolution of a PCC (upper panel) and a NCC (lower panel) instability. The PCC scenario is obtained with $\delta_{hj} = 4$ and $R = 0.5$ while the NCC scenario with $\delta_{hj} = -0.01$ and $R = 100$. These values are inspired from microemulsions data.

3 Experiments

As mentioned in the introduction, microemulsions show non-negligible cross-diffusion due to compartmentalization which can be characterised by means of the Taylor dispersion technique [2,3,14,15]. In a wide range of the ME composition the diffusion matrix D presents a large and positive cross-diffusivity D_{12} relating the motion of the water (species 1) to the flux of the surfactant (species 2) while a small negative cross-diffusion D_{21} quantifies the effect of water motion on AOT diffusion:

$$D = \begin{pmatrix} 0.6 & 7.8 \\ -0.01 & 1.3 \end{pmatrix} \times 10^{-6} \text{cm}^2 \text{ s}^{-1} \tag{10}$$

Thus, ME represent a good model to test the existence of PCC and NCC, depending on the initial gradient imposed either in the concentration of water or AOT.

The most convenient reactor for studying convective dynamics at the interface between two ME of different compositions consists of a vertically oriented Hele-Shaw cell, composed of two borosilicate glasses separated by a teflon spacer of 0.10 mm (see Fig. 5) [2,3,18]. Two different water-in-oil (W/O) reverse microemulsions were filled in the reactor through the inlet ports positioned at the top and the bottom of the cell. The excess of the solutions is pumped out through the cell's outlets until a flat interface between the two liquids is obtained. Each of the two microemulsions having different composition initially occupies half of the reactor height. The top and bottom solutions are prepared by using distilled

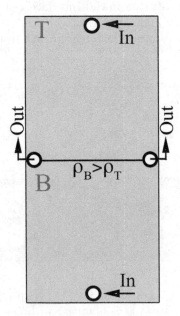

Fig. 5. Sketch of a Hele-Shaw cell.

water and a concentrated AOT in octane stock solution, conveniently diluted until the desired composition is reached. The dynamics of the interface is generally monitored by using a schlieren technique [17], which allows to observe the gradients in the refractive index between the two microemulsions, due to their density differences and, therefore, monitor the convective motions in solutions without the presence of dyes.

In all experiments the ME located in the bottom is denser than the one on the top, thus the system is initially stable towards buoyantly driven instabilities. Thanks to the presence of a negative and a positive cross-diffusion term, we can explore both the negative (NCC) and positive (PCC) cross-diffusive-driven instability by imposing a concentration jump in water or AOT, respectively. For the negative cross-diffusive case, theoretically described by the Fig. 3(d–f) and Fig. 4(b), the two ME are prepared as to have [AOT] initially constant everywhere, while the ME on the top has less water than that placed below. For a certain range of ΔH_2O, depending on the experimental conditions, the initially stable configuration becomes unstable after few minutes from the beginning of the experiment. As an example, Fig. 6(a) shows the appearance of convective vortices localized in the top and in the bottom layer at symmetric distances

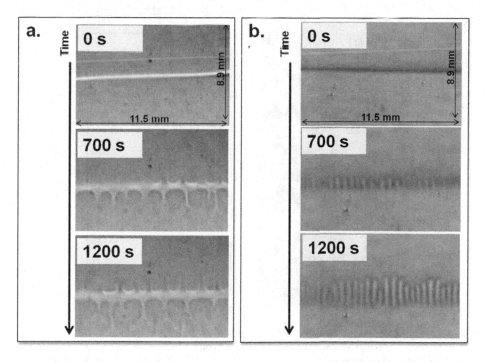

Fig. 6. Typical examples of cross-diffusion-driven instabilities obtained with microemulsions. (a) Three snapshots taken at t = 0, 700 and 1200 s displaying the evolution of the NCC mode obtained with $\Delta H_2O_{bottom-top} \simeq 1$ M. (b) The PCC scenario is obtained with $\Delta AOT_{bottom-top} \simeq 0.05$ M.

180

above and below the unperturbed interface, when $\Delta H_2O \simeq 1$ M. This compares favourably with the classification and the spatio-temporal dynamics shown in the theoretical section.

In order to induce a positive cross-diffusive-driven convective instability, theoretically described by the Fig. 3(a–c), the initial composition of the two ME has to be the other way around respect to the NCC case: $[H_2O]$ is constant throughout the upper and the lower layers, while $[AOT]$ is larger in the bottom ME.

Again, for a certain interval of $\Delta[AOT]$ the interface between the two ME can be destabilised, and in this case it deforms into fingers that grow vertically and symmetrically with time across the interface (see Fig. 6(b)). These structures are successfully predicted by the analysis of the density profile obtained from the cross-diffusion model in analogous conditions and the phenomenology favourably compares with nonlinear simulations.

As a proof of concept, the PCC scenario has also been investigated in microemulsions when a further solute is dissolved in the water core of the bottom solution. It was shown, in fact, that water-soluble molecules, when dissolved in ME, induce a large and positive co-flux both in the AOT and in the H_2O [15]. Figure 7 shows the appearance of convective fingers, which grow upwards and downwards around the interface, in a system where $KMnO_4$ has been added to

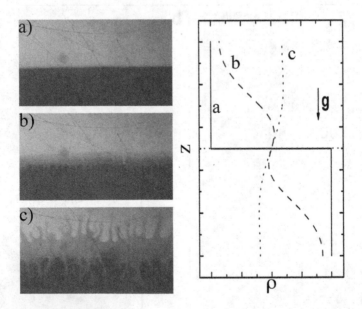

Fig. 7. Panels (a)–(c) show the spatio-temporal evolution of the interface between two stratified identical MEs, where the bottom one also contains the water soluble species $KMnO_4$. The snapshots are taken at 0, 300 and 600 s after the beginning of the experiment, respectively. The panel on the right describes the evolution of the density profile obtained by solving the diffusion Eqs. (1–2) for a 3-components system. The profiles a,b,c in the right panel describe the density distribution along the gravitational axis at the time corresponding to panels (a), (b) and (c), respectively.

the bottom ME. The composition of the two ME is such that no gradients are present in the concentrations of the water and the surfactant; in this configuration the interface is initially stable since the bottom solution is denser than the top one (Fig. 7(a)). However, the flux of $KMnO_4$ triggers the motion of a large quantity of H_2O and AOT molecules since both cross-diffusion terms are large and positive. Therefore, when $KMnO_4$ diffuses from the bottom to the upper layer, it drags along both H_2O and AOT molecules thus generating a non-monotonic density distribution around the contact line, destabilising the initially stable system (Fig. 7(b) and (c)). Even in this case, the evolution of the density profile has been reconstructed by integrating the Fickian Eqs. (1–2) for a 3-components system and combining it to the state Eq. (3). The resulting graph depicted in the right panel of Fig. 7 shows how the system undergoes a typical PCC scenario.

4   Conclusions

In this work we have presented a proof of concept that, by inducing convective transport, cross-diffusion can promote an efficient way to transport of spatio-temporal information in response to specific chemical gradients. We have shown how cross-diffusion-driven convection can be isolated and controlled in double-layer stratifications in the gravitational field. Furthermore, the resulting instability scenarios have been described in a general theoretical framework. Numerical results have been validated by means of experiments carried out with microemulsions which represent an ideal model system to investigate the two possible convective patterns, being characterized by a positive and a negative off-diagonal term in the cross-diffusion matrix.

Microemulsions also feature a convenient dispersed medium in this context because they have been widely studied in combination with chemical reactions. This paves the avenue to new chemo-hydrodynamic pattern formation and to unravel how cross-diffusion-driven convection interacts with linear and nonlinear chemical kinetics.

Future challenges are to engineer the results of this study to issues of applied interest, for instance in the realm of pollutant remediation processes and drug delivery.

Acknowledgments. F.R. was supported by the grant ORSA149477 funded by the University of Salerno (FARB ex 60 %) and gratefully acknowledges the support through the COST Action CM1304 (Emergence and Evolution of Complex Chemical Systems). M.A.B. gratefully acknowledges Regione Sardegna for financial support in the framework of "Asse IV Capitale Umano, Obiettivo Operativo 1.3 Linea di Attività 1.3.1 del P.O.R. Sardegna F.S.E. 2007/2013 - Progetti in forma associata e/o partenariale C.U.P. E85E12000060009" and the European Space Agency (ESA) Topical Team on "Chemo-Hydrodynamic Pattern Formation at Interfaces".

References

1. Budroni, M.A.: Cross-diffusion-driven hydrodynamic instabilities in a double-layer system: general classification and nonlinear simulations. Phys. Rev. E **92**(6), 063007 (2015)
2. Budroni, M.A., Carballido-Landeira, J., Intiso, A., De Wit, A., Rossi, F.: Interfacial hydrodynamic instabilities driven by cross-diffusion in reverse microemulsions. Chaos Interdisc. J. Nonlinear Sci. **25**(6), 064502 (2015)
3. Budroni, M.A., Lemaigre, L., De Wit, A., Rossi, F.: Cross-diffusion-induced convective patterns in microemulsion systems. Phys. Chem. Chem. Phys. **17**(3), 1593–1600 (2015)
4. Budroni, M.A., Rossi, F.: A novel mechanism for in situ nucleation of spirals controlled by the interplay between phase fronts and reaction diffusion waves in an oscillatory medium. J. Phys. Chem. C **119**(17), 9411–9417 (2015)
5. Carballido-Landeira, J., Trevelyan, P.M.J., Almarcha, C., De Wit, A.: Mixed-mode instability of a miscible interface due to coupling between Rayleigh-Taylor and double-diffusive convective modes. Phys. Fluids **25**(2), 024107 (2013)
6. Epstein, I.R.: Coupled chemical oscillators and emergent system properties. Chem. Commun. **50**(74), 10758–10767 (2014)
7. Epstein, I.R., Vanag, V.K., Balazs, A.C., Kuksenok, O., Dayal, P., Bhattacharya, A.: Chemical oscillators in structured media. Acc. Chem. Res. **45**(12), 2160–2168 (2011)
8. Ganguli, A.K., Ganguly, A., Vaidya, S.: Microemulsion-based synthesis of nanocrystalline materials. Chem. Soc. Rev. **39**(2), 474–485 (2010)
9. Leaist, D.G.: Relating multicomponent mutual diffusion and intradiffusion for massociating solutes. Application to coupled diffusion in water-in-oil microemulsions. Phys. Chem. Chem. Phys. **4**(19), 4732–4739 (2002)
10. Leaist, D., Hao, L.: Size distribution model for chemical interdiffusion in water AOT Heptane water-in-oil microemulsions. J. Phys. Chem. **99**(34), 12896–12901 (1995)
11. Rossi, F., Budroni, M.A., Marchettini, N., Carballido-Landeira, J.: Segmented waves in a reaction-diffusion-convection system. Chaos Interdisc. J. Nonlinear Sci. **22**(3), 037109 (2012)
12. Rossi, F., Liveri, M.L.T.: Chemical self-organization in self-assembling biomimetic systems. Ecol. Model. **220**(16), 1857–1864 (2009)
13. Rossi, F., Ristori, S., Marchettini, N., Pantani, O.L.: Functionalized clay microparticles as catalysts for chemical oscillators. J. Phys. Chem. C **118**(42), 24389–24396 (2014)
14. Rossi, F., Vanag, V.K., Epstein, I.R.: Pentanary cross-diffusion in water-in-oil microemulsions loaded with two components of the Belousov-Zhabotinsky reaction. Chem. Eur. J. **17**(7), 2138–2145 (2011)
15. Rossi, F., Vanag, V.K., Tiezzi, E., Epstein, I.R.: Quaternary cross-diffusion in water-in-oil microemulsions loaded with a component of the Belousov-Zhabotinsky reaction. J. Phys. Chem. B **114**(24), 8140–8146 (2010)
16. Rossi, F., Zenati, A., Ristori, S., Noel, J.M., Cabuil, V., Kanoufi, F., Abou-Hassan, A.: Activatory coupling among oscillating droplets produced in microfluidic based devices. Int. J. Unconventional Comput. **11**(1), 23–36 (2015)
17. Settles, G.S.: Schlieren and Shadowgraph Techniques. Springer, Berlin (2001)
18. Shi, Y., Eckert, K.: A novel Hele-Shaw cell design for the analysis of hydrodynamic instabilities in liquid systems. Chem. Eng. Sci. **63**(13), 3560–3563 (2008)

19. Taylor, A.F., Tinsley, M.R., Wang, F., Huang, Z., Showalter, K.: Dynamical quorum sensing and synchronization in large populations of chemical oscillators. Science **323**(5914), 614–617 (2009)
20. Tomasi, R., Noel, J.M., Zenati, A., Ristori, S., Rossi, F., Cabuil, V., Kanoufi, F., Abou-Hassan, A.: Chemical communication between liposomes encapsulating a chemical oscillatory reaction. Chem. Sci. **5**(5), 1854–1859 (2014)
21. Tompkins, N., Li, N., Girabawe, C., Heymann, M., Ermentrout, G.B., Epstein, I.R., Fraden, S.: Testing Turings theory of morphogenesis in chemical cells. Proc. Natl. Acad. Sci. **111**(12), 4397–4402 (2014)
22. Torbensen, K., Rossi, F., Pantani, O.L., Ristori, S., Abou-Hassan, A.: Interaction of the Belousov-Zhabotinsky reaction with phospholipid engineered membranes. J. Phys. Chem. B **119**(32), 10224–10230 (2015)
23. Trevelyan, P.M.J., Almarcha, C., De Wit, A.: Buoyancy-driven instabilities of miscible two-layer stratifications in porous media and Hele-Shaw cells. J. Fluid Mech. **670**, 38–65 (2011)
24. Vanag, V.K.: Waves and patterns in reaction-diffusion systems. Belousov-Zhabotinsky reaction in water-in-oil microemulsions. Physics-Uspekhi **47**(9), 923–941 (2004)
25. Vanag, V.K., Epstein, I.R.: Pattern formation in a tunable medium: the Belousov-Zhabotinsky reaction in an aerosol OT microemulsion. Phys. Rev. Lett. **87**(22), 228301–228304 (2001)
26. Vanag, V.K., Epstein, I.R.: Cross-diffusion and pattern formation in reaction-diffusion systems. Phys. Chem. Chem. Phys. **11**(6), 897–912 (2009)
27. Zemskov, E.P., Kassner, K., Hauser, M.J.B., Horsthemke, W.: Turing space in reaction-diffusion systems with density-dependent cross diffusion. Phys. Rev. E **87**(3), 032906 (2013)

Giant Vesicles as Compartmentalized Bio-reactors: A 3D Modelling Approach

Fabio Mavelli[1]([✉]), Emiliano Altamura[1], and Pasquale Stano[2]

[1] Chemistry Department, Aldo Moro University, Via Orabona 4, 70125 Bari, Italy
{fabio.mavelli,emiliano.altamura}@uniba.it
[2] Science Department, Roma Tre University, Viale G. Marconi 446, 00146 Rome, Italy
stano@uniroma3.it

Abstract. Giant lipid vesicles have been extensively used in recent years as in vitro artificial models for protocells, i.e. primitive cell models or synthetic cell-like systems of minimal complexity. Due to their dimensions in the micrometer range, giant vesicles can be designed as micro-sized enzymatic chemical reactors fed by a flux of substrates from the outside and monitored by confocal light microscopy in order to follow the production of fluorescence compounds. In this paper we present a 3D modelling approach to the simulation of giant vesicle where enzymatic reactions take place, and we apply this approach to a case study of a three-enzymes metabolic pathway.

1 Introduction

Giant lipid vesicles have been extensively used in recent years as in vitro artificial cell models embracing both origins-of-life studies [1–4] and modern synthetic biology [5–10]. This line of research is aimed at designing, constructing, and characterizing micro-compartmentalized structures of minimal complexity (protocells) that share with primitive cells or with modern living cells their peculiar static and dynamic organization. More recently, lipid vesicels have been also proposed for studying nonlinear chemical reactions and chemical communication among compartments [11–13].

Several experimental approaches are currently applied in this field and most of them rely on lipid vesicles (both from fatty acids or phospholipids), but polymer vesicles [14] and coacervates have been also used [15]. Giant lipid vesicles (here called GVs) [16] are used for constructing cell-like systems since enzymes and other macromolecules are easily encapsulated inside them [2], allowing the construction of micro-compartmentalized systems capable of programmable behaviors [10]. Moreover, since GVs are micro-sized compartments, their size allows using traditional light microscopy to directly follow each giant vesicle. This is a great advantage when compared with electron-microscopy. GVs can be analyzed while they are functional in liquid water, while (bio)chemical reactions are running in their lumen, and all chemical and morphological transformations can be followed in real-time. Confocal light microscopy allows monitoring of individual GVs, provided that fluorescent probes are used, so that the variations of the internal fluorescence can be related to the production/consumption of internal compounds.

© Springer International Publishing Switzerland 2016
F. Rossi et al. (Eds.): WIVACE 2015, CCIS 587, pp. 184–196, 2016.
DOI: 10.1007/978-3-319-32695-5_17

While several experimental advancements have been reported recently, less attention has been paid to numerical modeling. This sharply contrast with the integrative approach supported by synthetic biology, the bioengineering discipline that study, among other subjects, artificial (synthetic) cells. Synthetic biology aims at understanding biological systems by a constructive approach, characterizing all processes in quantitative manner. Enzyme-containing vesicles, due to their limited complexity, are excellent systems for combining experiments and simulations. Modelling these supramolecular reacting systems is of great interest both to better understand the dynamics of enzymatic reactions in confined space, but also to improve the design and the implementation of new metabolic micro-reactors.

In previous theoretical works [17–21] we studied the time behavior of nano-sized reacting vesicles by using a stochastic approach in order to elucidate the role of intrinsic fluctuations, i.e. fluctuations in the reactions occurring time (*intrinsic stochasticity*). On the other hand, two deterministic approaches are instead suitable for modelling enzymatic pathways taking place in micrometer-sized GVs: the 0D and the 3D approach respectively. Both assume that intrinsic noise is negligible in first approximation and that random fluctuations observed in the time behavior of a vesicle population are mainly due to *extrinsic stochasticity*, i.e., due to the different sizes and different composition of the reacting compartments as shown by Fig. 1.

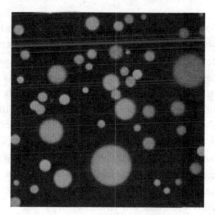

Fig. 1. Confocal light microscopy image of an aqueous suspension of POPC giant vesicles filled by calcein, and prepared by the phase transfer method. The preparation procedure gives a very poly-dispersed vesicle solution both in size of lipid compartments and concentration of encapsulated solutes as shown by the different fluorescence intensity of the vesicle cores.

Even though these two approaches share this common assumption, 0D and 3D models differ in the description level and in the modelling purposes.

The 0D modelling aims to describe the average time behavior of reacting GV population taking into account the size dispersion and the solute concentration distribution. This approach allows to elucidate the extrinsic stochastic effects of different random distribution of entrapped biomolecules on the time behavior of the vesicle suspension. In fact, it has been recently reported that the Poisson and Gaussian distributions, in some cases, cannot account for the experimentally observed solute distribution in the vesicle population [22, 23] and, based on these experimental observations, we have developed a 0D model for the translations-transcription machinery taking place inside lipid vesicles in order to test the Power Law as a more suitable solute distribution [24].

Contrariwise, the 3D approach describes individual GVs giving also morphological 3D-space details and allows to take into account explicitly the diffusion of substrates, through the external solution and in the internal vesicle water core, along with the molecular transport across the lipid membrane. The theoretical outcomes can be contrasted with confocal microscopy analysis and can be useful in designing ongoing communication experiments among GVs or GVs and living cells [25].

This paper is focused on the deterministic 3D modelling approach that will be applied to a three-enzymes metabolic pathway as a case study. In Sect. 2, the three-enzymes metabolic pathway will be introduced and its time evolution in a bulk solution will be reproduced by a simple kinetic model. In Sect. 3, different possible scenarios of compartmentalization will presented. The 3D model will be illustrated in its computational details in Sect. 4, while the computational results will be discussed in Sect. 5. Finally, some conclusion will be drawn in the last section.

2   The 3-Enzymes Metabolic Pathway

Figure 2 reports the 3-enzymes metabolic pathway that is proposed as a case study for this work. It consists in a branched enzymatic network for the oxidation of 2'-7'dichloro-dihydro-fluorescein diacetate (DCFH$_2$-DA) to 2'-7'dichloro-fluorescein (DFC), a fluorescent compound that can be followed fluorometrically. In this simplified metabolism three different enzymes are involved: glucose oxidase (GOX) that converts glucose into hydrogen peroxidase and D-gluconolactone (not shown in the scheme of Fig. 2); Carbonic Anhydrases (CA) that catalyses the rapid interconversion of DCFH$_2$-DA in 2'-7'dichloro-dihydro-fluorescein (DCFH$_2$) and finally Horseradish Peroxidase (HRP) that oxidises the intermediate DCFH$_2$ to DFC thanks to H$_2$O$_2$ produced by GOX. Figure 2 reports also the reaction rate expressions for the three enzymatic steps along with the kinetic parameters that will be used in this work as derived by the BRENDA enzyme database (http://www.brenda-enzymes.org/).

Figure 3 shows the comparison between the experimental time course of the DFC concentration produced by the enzymatic pathway in a bulk aqueous solution (red points) and the theoretical curve obtained by solving the kinetic ODE set. Both the curves have been normalized dividing each data series for the respective value at time 1000 s.

Although the match between data point and the theoretical curve could be improved (especially for long time values), the bulk kinetic model well reproduces the time scale of the [DFC] evolution in the time range of interest. This validates the enzymatic rate expressions and the kinetic parameters reported in Fig. 2, and therefore they will be used in the 3D model that will be described in the next sections.

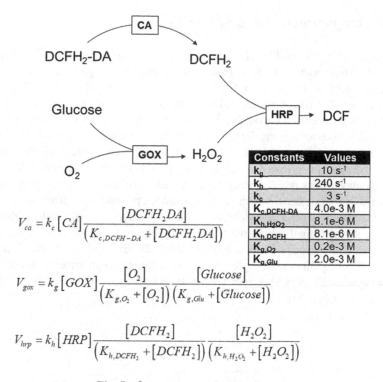

Fig. 2. 3-enzymes metabolic pathways

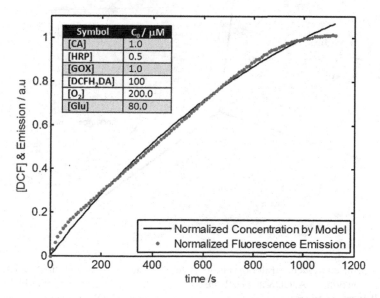

Fig. 3. Comparison between the time course of the experimental production of DCF performed in a bulk aqueous solution and the theoretical model. In the upper left corner, the table with the initial concentrations of the three enzymes and substrates is reported (Color figure online).

3   The Compartmentalized Scenarios

In presence of lipid vesicles, different scenarios con be envisaged for the implementation of the 3-enzymes metabolic pathway depending on where the enzymes are located. These scenarios are sketched in Fig. 4: the first one is the case where all the three enzymes are encapsulated in the same compartment. Therefore, the two substrates glucose and DCFH$_2$-DA must freely diffuse from the outside across the lipid membrane into the internal core of the vesicle where they are chemically converted by the enzymes. Compartments so designed could act as fluorescent sensors for the presence of glucose in the external solution, although the glucose permeability is lower compared to other less hydrophilic molecules. The second scenario circumvent this problem since the enzymatic conversion of glucose by the GOX takes place in the external environment and the hydrogen peroxides can freely diffuse across the lipid membrane, with a permeability which is 3 orders of magnitude higher when compared to those of small organic molecules [26]. This kind of compartments can be seen as fluorescent sensors for H$_2$O$_2$. Finally in the third scenario, the two branches of the 3-enzymes metabolic pathway are segregated in two different compartments that communicates by a signal molecule, the hydrogen peroxide, although the problem of the low glucose permeability is not avoided.

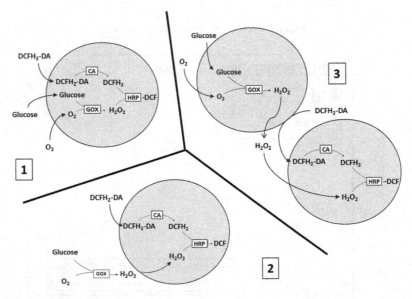

Fig. 4. Three different scenarios for the encapsulation the three enzymes in giant vesicles. In the first, the three enzymes CA, GOX and HRP are co-encapsulated within the vesicles. In the second, CA and HRP are co-encapsulated within the vesicles, whereas GOX is present in the external solution. In the third, CA and HRP are co-encapsulated within a vesicle subpopulation, whereas GOX is encapsulated within another vesicle subpopulation; the two populations, prepared separately, are then mixed.

In this paper the attention will be focused only on scenarios 2 and 3 since we are mainly interested in designing and optimizing giant vesicles for the experimental implementation of sort of a chemical communication between reacting compartments.

4   Giant Vesicles 3D Model: Computational Details

This approach is a deterministic approach where molecular diffusion and molecular transportation across the lipid membrane are explicitly considered. Therefore, the concentration time evolution of each species in the reacting system is described by a function depending both on time and on space coordinates: $C_i(t, x, y, z)$, C_i being the concentration of i-th species at time t in the space point (x, y, z). The global reacting system, i.e. the vesicle suspension, is reduced to a cubic box containing only one vesicle (scenario 1 or 2) or a rectangular parallelepiped containing at least two vesicles (scenario 3) as it is shown in Fig. 5.

Periodic boundary condition are applied to the box walls:

$$C_i\left(t, -L_x/2, y, z\right) = C_i\left(t, L_x/2, y, z\right)$$

$$\mathbf{n}\left(-L_x/2, y, z\right)\nabla C_i\left(t, -L_x/2, y, z\right) = \mathbf{n}\left(L_x/2, y, z\right)\nabla C_i\left(t, L_x/2, y, z\right)$$

$$C_i\left(t, x, -L_y/2, z\right) = C_i\left(t, x, L_y/2, z\right)$$

$$\mathbf{n}\left(x, -L_y/2, z\right)\nabla C_i\left(t, -L_y/2, y, z\right) = \mathbf{n}\left(x, L_y/2, z\right)\nabla C_i\left(t, x, L_y/2, z\right)$$

$$C_i\left(t, x, y, -L_z/2\right) = C_i\left(t, x, y, L_z/2\right)$$

$$\mathbf{n}\left(x, y, -L_z/2\right)\nabla C_i\left(t, x, y, -L_z/2\right) = \mathbf{n}\left(x, y, L_z/2\right)\nabla C_i\left(t, x, y, L_z/2\right) \tag{1}$$

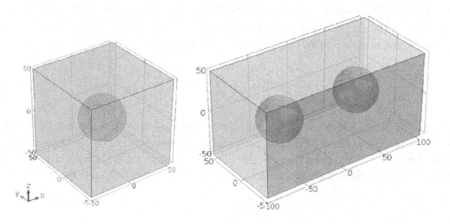

Fig. 5. Schematic representation of a vesicle suspension according to the scenarios 1 and 2 on the left, and scenario 3 on the right.

where **n** is the surface unit vector and L_z, L_y, and L_x, are the lengths of the box sides. For a cubic box $L = L_z = L_y = L_x$ while for a rectangular parallelepiped $L = L_z = L_y \neq L_x = 2L$ respectively. The size of the box can be defined by knowing the lipid concentration [*Lip*] and the vesicle aggregation number N_{agg}: $L = (N_{agg}/(N_A[Lip]))^{1/3}$, N_A being the Avogadro's Number. In this paper giant spherical vesicles with a 25 μm radius are always considered and each of them is contained in squared box of $L = 100$ μm. This corresponds to a vesicle concentration of 1.7e−15 M and a lipid concentration of 35 μm assuming a monodispersed vesicle population and a lipid head area of 0.72 nm^2.

Table 1. Permeability [27] and Diffusion coefficients [28] for different substrates and metabolites. The permeability of DCFH$_2$ and DCF have been set zero since both these molecules dissociate in water solution in the working conditions (pH around 7.0) giving negatively charged compounds.

Permeability [cm/s]	Diffusion Coefficient [cm^2/s]	Compound
0.5e-7	6.0E-06	Glucose
1e-6	6.0E-06	DCFH$_2$-DA
1e-3	1.8e-04	H$_2$O$_2$
0.0	6.0E-06	DCFH$_2$
0.0	6.0E-06	DCF

The space in the box is then decomposed in different reacting domains, for instance the vesicle core and the external solution, and free diffusion is allowed for all species in each system domains:

$$\frac{\partial C_i^\delta}{\partial t} = D_i \nabla C_i^\delta \tag{2}$$

where subscript i indicates the species while superscript δ the space domain and D_i is the diffusion coefficient of the i-*th* species. If in the δ-*th* domain the i-*th* species is involved in metabolic reactions the partial differential Eq. (2) becomes

$$\frac{\partial C_i^\delta}{\partial t} = \nabla^2 C_i^\delta + \sum_R \alpha_i^R V_R \tag{3}$$

where α_i^R is the stochiometric coefficent of the i-*th* species in the R-*th* reaction taken negative for reactants, positive for products and null if the species is not inolved in the reactive step. Across the GVs' boundary, i.e. the lipid membrane, passive transport takes place according to the molecular permeability $\wp_i$ of species along the normal to the surface **n**:

$$\mathbf{n}\nabla C_i = \wp_i \left(C_i^{Ex} - C_i^{In} \right) \tag{4}$$

In Table 1, the permeability and the diffusion coefficients for the different substrates and metabolites are reported along with the literature references, while in Table 2 the initial concentrations of substrates and enzymes are listed. The PDE set as been numerically solved by using the COMSOL Multiphysics Software.

Table 2. Initial concentrations of substrates and enzymes.

Species	C_0 (μM)
[CA]	1.0
[HRP]	1.0
[GOX]	0.5
[DCFH$_2$DA]	100.0
[O$_2$]	200.0
[Glucose]	20.0

5   Results

In this section some preliminary results of the numerical solution of the PDE set will be presented only for the two of the compartmentalized enzymatic reacting systems previously described: scenario 2 and 3 respectively.

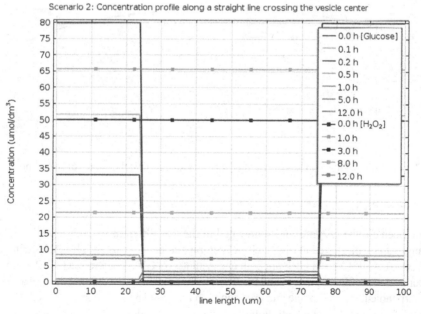

Fig. 6. Scenario 2: Glucose (line) and Hydrogen Peroxides (line and dots) concentration profiles along a horizontal line perpendicular to ZY plane and crossing the giant vesicle in its center.

5.1   Scenario 2

In Fig. 6, the profiles of the glucose and hydrogen peroxide concentration calculated along a straight line perpendicular to the ZY plane and crossing the giant vesicle center are reported in time. Although after one hour the external glucose is almost exhausted and the produced hydrogen peroxide quickly diffuse across the membrane, more than 12 h are necessary for the total conversion of $DCFH_2$-CA into DCF. The latter reaches the limit concentration $[Glucose]_0(V_{box}-V_{GV})/V_{GV} = 1.14$ mM, see Fig. 7. The plot in Fig. 6 also shows that a residual amount of glucose remains entrapped into the vesicles and sustains the DCF production in the long time diffusing back in the external environment where it is in turn converted in H_2O_2.

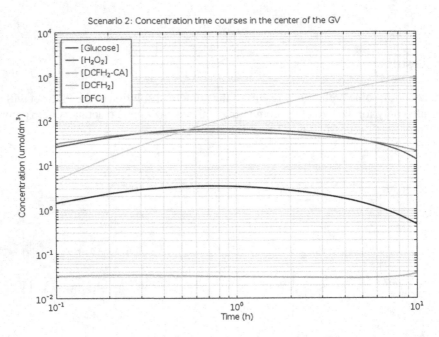

Fig. 7.   Scenario 2:Time courses of reacting species calculated in the center of the giant vesicles

5.2   Scenario 3

In this Scenario the enzymes are segregated in the two different vesicles contained in the rectangular box illustrated in Fig. 5. GOX is entrapped in the giant vesicle GV1, centered in $(-L/2, 0, 0)$, while CA and HRP are both encapsulated in GV2, centered in $(+L/2, 0, 0)$. The concentration profile of Glucose and the fluorescent product DCF calculated along a straight line perpendicular to YZ plane and crossing both the vesicles centers are reported in Fig. 8. After 12 h, the external concentration of glucose is poorly decreased going from 80 μM to 72 μM while the internal concentration of DCF in GV2 reach 190 μM, a concentration value about six time lower than those in the previous scenario. This can be ascribed to the fact that

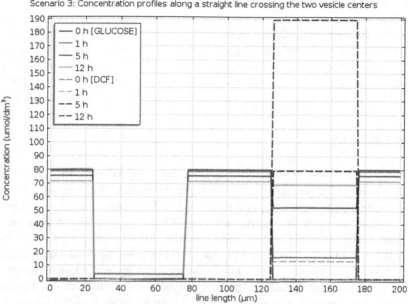

Fig. 8. Scenario 3: Glucose (line) and DCF (dashed line) concentration profiles along a horizontal line perpendicular to ZY plane and crossing the center of the two giant vesicles.

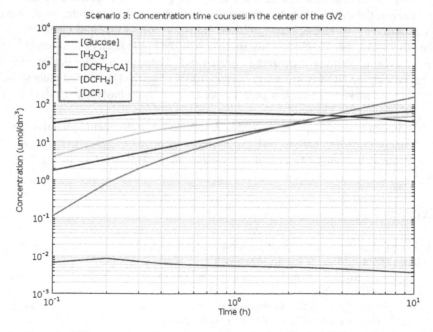

Fig. 9. Scenario 3:Time courses of reacting species calculated in the center of the giant vesicle GV2 located in $(+L/2, 0, 0)$.

in order to be enzymatically converted glucose must diffuse across the lipid membrane and this slow down the production of hydrogen peroxide.

In fact, conversely from the previous scenario, the concentration versus time courses calculated in the center of GV2 and reported in Fig. 9 show that now the level of hydrogen peroxide inside this vesicle remains very low throughout the process while the concentration of the intermediate $DCFH_2$ is comparable to those of the substrate $DCFH_2$-CA. This is due to the slow enzymatic conversion of the intermediate into the product DFC catalyzed by HRP, because of the low concentration of H_2O_2 slowly produced by GV1.

6   Conclusions

In this work we presented a 3D deterministic modelling approach to the study of enzymatic metabolic pathways taking place inside giant vesicles. This approach is focused on reproducing the time behavior of a single or few GVs seen as micro-sized reactors fed by spontaneous substrate transportation through the vesicle lipid membrane from the outside, taking also into account the molecular diffusion of species driven by the concentration gradients. The approach has been applied to 3-enzymes metabolic pathway taken as a case study in two different scenarios by using kinetic parameters, diffusion coefficients and membrane permeability derived from literature. Even if the presented results are preliminary and a better comparison with experimental data is necessary to deeply validate the enzymatic model, the presented approach clearly shows a new way of simulating reacting giant vesicles that can be greatly useful in designing and optimizing compartmentalized reacting systems that mimic cellular behavior. In particular, this theoretical study has shown that molecular diffusion in the solution bulk is faster than reaction kinetics and molecular transport across the lipid membrane and, on the other hand, that the glucose transportation is the bottle neck of the compartmentalized metabolic pathway according to Scenario 3. Therefore a better design of giant vesicles entrapping GOX is necessary in order to increase the flux rate of glucose from the outside and make these compartments more efficient supramolecular devices for the hydrogen peroxide production.

More in general, our approach would be integrated in experiment/simulation cycles for optimization of biomimetic cell-like systems of different complexity, especially when chemical exchanges from and to vesicles are key parameters to regulate and control.

References

1. Monnard, P.-A., Deamer, D.W.: Membrane self-assembly processes: steps toward the first cellular life. Anat. Rec. **268**, 196–207 (2002)
2. Luisi, P.L., Ferri, F., Stano, P.: Approaches to semi-synthetic minimal cells: a review. Naturwissenschaften **93**, 1–13 (2006)
3. Mansy, S.S., Szostak, J.W.: Reconstructing the emergence of cellular life through the synthesis of model protocells. Cold Spring Harb. Symp. Quant. Biol. **74**, 47–54 (2009)

4. Kurihara, K., Okura, Y., Matsuo, M., Toyota, T., Suzuki, K., Sugawara, T.: A recursive vesicle-based model protocell with a primitive model cell cycle. Nat. Commun. **6**, 8352 (2015). doi:10.1038/ncomms9352

5. Yu, W., Sato, K., Wakabayashi, M., Nakaishi, T., Ko-Mitamura, F.P., Shima, Y., Urabe, I., Yomo, T.: Synthesis of functional protein in liposome. J. Biosci. Bioeng. **92**, 590–593 (2001)

6. Noireaux, V., Libchaber, A.: A vesicle bioreactor as a step toward an artificial cell assembly. Proc. Natl. Acad. Sci. U.S.A. **101**, 17669–17674 (2004)

7. Stano, P., Carrara, P., Kuruma, Y., Souza, T.P., Luisi, P.L.: Compartmentalized reactions as a case of soft-matter biotechnology: Synthesis of proteins and nucleic acids inside lipid vesicles. J. Mater. Chem. **21**, 18887–18902 (2011)

8. Nourian, Z., Danelon, C.: Linking genotype and phenotype in protein synthesizing liposomes with external supply of resources. ACS Synth. Biol. **2**, 186–193 (2013)

9. Stano, P., D'Aguanno, E., Bolz, J., Fahr, A., Luisi, P.L.: A remarkable self-organization process as the origin of primitive functional cells. Angew. Chem. Int. Ed. **52**, 13397–13400 (2013)

10. Lentini, R., Santero, S.P., Chizzolini, F., Cecchi, D., Fontana, J., Marchioretto, M., del Bianco, C., Terrell, J.L., Spencer, A.C., Martini, L., et al.: Integrating artificial with natural cells to translate chemical messages that direct E. coli behaviour. Nat. Commun. **5**, 4012 (2014). doi: 10.1038/ncomms5012

11. Tomasi, R., Noel, J.-M., Zenati, A., Ristori, S., Rossi, F., Cabuil, V., Kanoufi, F., Abou-Hassan, A.: Chemical communication between liposomes encapsulating a chemical oscillatory reaction. Chem. Sci. **5**(5), 1854–1859 (2014)

12. Stano, P., Wodlei, F., Carrara, P., Ristori, S., Marchettini, N., Rossi, F.: Approaches to molecular communication between synthetic compartments based on encapsulated chemical oscillators. In: Pizzuti, C., Spezzano, G. (eds.) WIVACE 2014. CCIS, vol. 445, pp. 58–74. Springer, Heidelberg (2014)

13. Rossi, F., Zenati, A., Ristori, S., Noel, J.-M., Cabuil, V., Kanoufi, F., Abou-Hassan, A.: Activatory coupling among oscillating droplets produced in microfluidic based devices. Int. J. Unconv. Comp. **11**(1), 23–36 (2015)

14. Martino, C., Kim, S.-H., Horsfall, L., Abbaspourrad, A., Rosser, S.J., Cooper, J., Weitz, D.A.: Protein expression, aggregation, and triggered release from polymersomes as artificial cell-like structures. Angew. Chem. Int. Ed. **51**, 6416–6420 (2012)

15. Tang, T.-Y.D., van Swaay, D., deMello, A., Anderson, J.L.R., Mann, S.: In vitro gene expression within membrane-free coacervate protocells. Chem. Commun. **51**, 11429–11432 (2015)

16. Walde, P., Cosentino, K., Engel, H., Stano, P.: Giant vesicles: preparations and applications. ChemBioChem **11**, 848–865 (2011)

17. Mavelli, F., Altamura, E., Cassidei, L., Stano, P.: Recent theoretical approaches to minimal artificial cells. Entropy **16**, 2488 (2014)

18. Shirt-Ediss, B., Ruiz-Mirazo, K., Mavelli, F., Sole, R.V.: Modelling lipid competition dynamics in heterogeneous protocell populations. Sci. Rep. **4**, 5675 (2014)

19. Mavelli, F., Ruiz-Mirazo, K.: Theoretical conditions for the stationary reproduction of model protocells. Integr. Biol. **5**, 324–341 (2013)

20. Mavelli, F.: Stochastic simulations of minimal cells: the Ribocell model. BMC Bioinform. **13**(4), S10 (2012)

21. Mavelli, F., Ruiz-Mirazo, K.: ENVIRONMENT: a computational platform to stochastically simulate reacting and self reproducing compartments. Phys. Biol. **3**, 36002 (2010)

22. Luisi, P.L., Allegretti, M., Souza, T., Steineger, F., Fahr, A., Stano, P.: Spontaneous protein crowding in liposomes: a new vista for the origin of cellular metabolism. ChemBioChem **11**, 1989 (2010)
23. Souza, T.P., Fahr, A., Luisi, P.L., Stano, P.: Spontaneous encapsulation and concentration of biological macromolecules in liposomes: an intriguing phenomenon and its relevance in origins of life. J. Mol. Evol. **79**, 179 (2014)
24. Mavelli, F., Stano, P.: Experiments and numerical modelling on the capture and concentration of transcription-translation machinery inside vesicles. Artif. Life **21**, 1–19 (2015)
25. Mavelli, F., Rampioni, G., Damiano, L., Messina, M., Leoni, L., Stano, P.: Molecular communication technology: general considerations on the use of synthetic cells and some hints from in silico modeling. In: Pizzuti, C., Spezzano, G. (eds.) WIVACE 2014. CCIS, vol. 445, pp. 169–189. Springer, Heidelberg (2014)
26. Bienert, G.P., Schjoerring, J.K., Jahn, T.P.: Membrane transport of hydrogen peroxide. Biochim. Biophys. Acta **1758**, 994–1003 (2006)
27. Robertson, R.N.: The Lively Membranes. Cambridge University Press, Cambridge (1983)
28. Daniels, F., Alberty, R.A.: Physical Chemistry. Wiley, New York (1961)

Engineering Enzyme-Driven Dynamic Behaviour in Lipid Vesicles

Ylenia Miele[1], Tamás Bánsági Jr.[2], Annette F. Taylor[2], Pasquale Stano[3], and Federico Rossi[1](✉)

[1] Department of Chemistry and Biology, University of Salerno,
Via Giovanni Paolo II 132, 84084 Fisciano (SA), Italy
frossi@unisa.it

[2] Department of Chemical and Biological Engineering, University of Sheffield,
Mappin Street, S1 3JD, Sheffield, UK

[3] Science Department, Roma Tre University, V.le Marconi 446, 00146 Rome, Italy

Abstract. The urea–urease system is a pH dependent enzymatic reaction that was proposed as a convenient model to study pH oscillations *in vitro*; here, in order to determine the best conditions for oscillations, a two-variable model is used in which acid and substrate, urea, are supplied at rates k_h and k_s from an external medium to an enzyme-containing compartment. Oscillations were observed between pH 4 and 8. Thus the reaction appears a good candidate for the observation of oscillations in experiments, providing the necessary condition that $k_h > k_s$ is met. In order to match these conditions, we devised an experimental system where we can ensure the fast transport of acid to the encapsulated urease, compared to that of urea. In particular, by means of the *droplet transfer method*, we encapsulate the enzyme, together with a suitable pH indicator, in a 1-palmitoyl-2-oleoyl-*sn*-glycero-3-phosphatidylcholine (POPC) lipid membrane, where differential diffusion of H^+ and urea is ensured by the different permeability (P_m) of membranes to the two species. Here we present preliminary tests for the stability of the enzymatic reaction in the presence of lipids and also the successful encapsulation of the enzyme into lipid vesicles.

Keywords: Enzymatic oscillators · urea–urease reaction · lipid vesicles · pH oscillators

1 Introduction

Chemical oscillations are a genuine manifestation of emergent complex behaviour taking place in far-from-equilibrium nonlinear chemical systems. Despite the fact that systems exhibiting oscillations in the concentrations of some of the reaction intermediates are known since the beginning of the past century, systematic and thorough study of oscillatory mechanisms and dynamics begun following the discovery of the Belousov-Zhabotinsky reaction, the most studied prototypical chemical oscillator [1]. The pH oscillators, along with the bromate and chlorite

© Springer International Publishing Switzerland 2016
F. Rossi et al. (Eds.): WIVACE 2015, CCIS 587, pp. 197–208, 2016.
DOI: 10.1007/978-3-319-32695-5_18

oscillator families, are the most employed to generate novel nonlinear temporal and spatial phenomena. Because of the ubiquity of the hydrogen ion in chemical and biological processes, pH oscillators offer the greatest promise for practical applications [2].

Enzyme catalysed reactions have been proposed as good candidates to obtain large amplitude pH oscillations [3]; in fact, due to the strong amphoteric character of enzymes, resulting from containing a large number of acid and basic groups, one of the most important factors affecting their activity is pH. Among the several enzymatic reactions known to depend on the pH, one candidate, the urea–urease, was recently revisited by us [4,5].

The urea–urease reaction is an enzyme-catalysed hydrolysis of urea, that produces ammonia and carbon dioxide according to the overall stoichiometry (1). This reaction occurs in numerous cellular systems, for example it is used by bacteria *H. pylori* in order to raise the local pH to protect itself from the harsh acidic environment of the stomach [6].

$$CO(NH_2)_2 + H_2O \xrightarrow{urease} 2NH_3 + CO_2 \tag{1}$$

The urea–urease reaction follows Michaelis–Menten kinetics and has a bell-shaped rate-pH curve with maximum at pH 7 (see Fig. 1). In non–buffered condition, the rate–pH curve can be exploited to obtain feedback-driven behaviour, for instance, through external stimuli, *e.g.* by delivering an acid or a base to the solution a reaction acceleration or inhibition can be obtained. Therefore, the conditions for spontaneous oscillations between two pH states can be achieved. This reaction has also the distinct advantages of high solubility and stability of substrate and enzyme in water making it suitable for experiments *in vitro*.

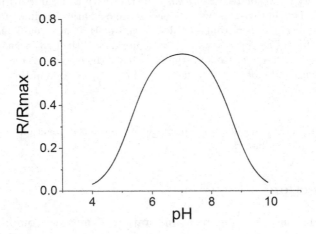

Fig. 1. Typical enzyme-catalysed rate R relative to maximum rate R_{max} as a function of pH.

The pH-dependency of the reaction rate has been proposed as key the feedback mechanism to pH oscillations obtained in a computational model involving a membrane-bound enzyme, papain, and the diffusion of substrate, an ester, from the surrounding solution. In the model, the diffusion coefficient of acid was five times larger than that of the other species [7]. Whilst the chemical nature of the feedback has been analysed, the importance of the inherent differential transport was perhaps under-emphasised in earlier work. Actually, differential transport arises naturally in cells as a result of variations in the permeability of the cell membrane to different species. For example, the transport of urea and other species into *H. pylori* involves proton-gated membrane channels [6]. The combination of an autocatalytic process and differential transport was already suggested to give rise to stationary patterns and oscillations in 1952 in the seminal theoretical work of Turing on the chemical basis of morphogenesis [8]. In his coupled cell model, the transport of the autocatalytic species was slower than that of the other species: this condition for pattern formation is often referred to as long range inhibition, short range activation. This idea has already been investigated in several nonlinear chemical systems confined or encapsulated in membranes [9–13], micro- or macro-emulsions [14–17], micelles [18–25] and organic or inorganic beads [26, 27].

In order to determine the conditions for oscillations in the urea–urease reaction, a two–variable reduced model was recently derived in which acid and substrate, urea, are supplied at rates k_h and k_s from an external medium to an enzyme-containing cell [28]. Oscillations were found between pH 4 and 10, thus the reaction appears a good candidate for the observation of oscillations in experiments, providing the necessary condition that $k_h > k_s$ is met. To test this hypothesis, the urease was immobilised in an alginate gel bead and placed the bead in a solution of urea and acid [29]. Bistability and hysteresis were obtained between low and high pH states. However, oscillations were not observed. It seems likely that this is because the diffusion of acid in the gel matrix is not significantly enhanced compared to that of the other species. Thus, we require a medium where we can ensure the fast transport of acid to the encapsulated urease compared to that of urea. One means of achieving this is by use of a 1-palmitoyl-2-oleoylphosphatidylcholine (POPC) lipid membrane, where differential diffusion of H^+ and urea is ensured by the different permeability (P_m) of the membrane to the two species ($P_m \sim 10^{-3}$ cm/s for H^+ and $\sim 10^{-6}$ cm/s for urea) [30, 31]. The successful encapsulation of the enzyme can be attained by means of the *droplet transfer method*, introduced by the Weitz group [32] and recently optimised by Luisi and collaborators [33, 34]. The method has been already tested in preliminary experiments on the confinement of the Belousov-Zhabotinsky chemical oscillator in a network of vesicles [35] and for the encapsulation of enzymatic reactions [36].

In order to follow the reaction in real-time, we devised an experimental system in which the enzyme is encapsulated, together with a fluorescence probe, into giant liposomes (diameter larger than $10\,\mu$m); in this way, the reaction dynamics over time can be easily monitored by an optical microscope equipped with a fluorescence module. Pyranine (8-hydroxy -1,3,6-pyrenetrisulfonate), has been chosen as the fluorescent probe, since its emission intensity (at 510 nm) is strongly dependent upon the pH of the solution, over the range 6–10. Pyranine has also the great advantage of not leaking out of the vesicles once entrapped therein [37]. Finally, in order to feed the encapsulated enzyme, liposomes will be dispersed in a solution containing urea and a suitable acid.

In Sect. 2.1 of this paper, we present the numerical simulation of the system where the two–variable model of the reaction is adapted for the lipid vesicles scenario. Finally, in Sect. 2.2 we show preliminary tests for the stability of the enzymatic reaction in the presence of lipids and also the successful encapsulation of the enzyme into lipid vesicles.

2   Results

2.1   Modeling Oscillatory Dynamics in Vesicles

The temporal dynamic of the urea–urease reaction in membranes in contact with a solution of urea (S) and acid (H^+), as depicted in Fig. 2, can be captured with the model (2):

$$
\frac{d[S]}{dt} = k_S([S]_0 - [S]) - R
$$

$$
\frac{d[H^+]}{dt} = \left(k_H \left([H^+]_0 - \frac{K_w}{[H^+]_0} - [H^+] + \frac{K_w}{[H^+]} \right) - 2R \right) \left(1 + \frac{K_w}{[H^+]^2} \right)^{-1} \tag{2}
$$

where

$$
R = \frac{k_E[E][S]}{(K_m + [S]) \left(1 + \frac{K_{ES2}}{[H^+]} + \frac{[H^+]}{K_{ES1}} \right)}
$$

$k_E = 3.7 \times 10^{-6}\,\mathrm{mL\,M\,u^{-1}s^{-1}}$, $K_m = 3 \times 10^{-3}\,\mathrm{M}$, $K_{ES1} = 5 \times 10^{-6}\,\mathrm{M}$, $K_{ES2} = 2 \times 10^{-9}\,\mathrm{M}$ are urease specific quantities and $K_w = 10^{-14}\,\mathrm{M^2}$ is the ion product of water. In this two–variable model, simplified from an 8-variable model previously used for studying bistability [4] and front propagation [5], the transport of substrate and acid are approximated as a simple exchange of matter through the boundary between a membrane or particle and the reservoir (with bulk concentrations $[S]_0$, $[H^+]_0$) governed by coefficients k_S and k_H, respectively. For our vesicles, they were calculated as $k_i = 6P_i/D$ where P_i denotes the permeability coefficients ($P_S = 2.45 \times 10^{-6}\,\mathrm{cm/s}$ and $P_H = 1.82 \times 10^{-3}\,\mathrm{cm/s}$ for bilayers made of lipids having a 16 carbon long hydrocarbon chain [31]) and $D = 10\,\mu\mathrm{m}$ is the diameter [30]. Equation (2) was scaled (See: Appendix) for analysis carried out with XPPAUT [38]. The modeling results are summarized in Fig. 3.

The model exhibits bistability and oscillations in a wide range of parameters. It is worthwhile to note, however, that the amplitude of oscillations (Fig. 3a inset) exceeds those obtained in the full model (not shown). The presence of carbon dioxide and its derivatives reduces the range of oscillations to about pH 4–6 with negligible effect on the frequency, hence pH indicators for monitoring oscillations

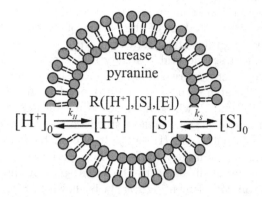

Fig. 2. Lipid vesicle containing the urea–urease reaction and the fluorescence probe: R = reaction rate, k_H = exchange rate of the acid, k_S = exchange rate of the urea, S, with the external solution of concentrations $[H^+]_0$ and $[S]_0$.

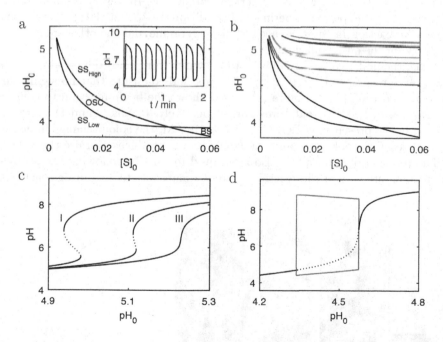

Fig. 3. (a) Phase diagram and oscillations for E = 1300 u/mL; (b) Phase diagrams for E = 1300 (–), 50 (–), 25 ( ), 15 (–), 5 (–) u/mL; (c) Bistability for $[S]_0$ = 0.15 M: E = 18 (I), 12 (II), 9 (III) u/mL; (d) Bifurcation diagram for E = 1300 u/mL, $[S]_0$ = 0.01 M, gray line indicates the size of limitcycles; (c–d) Solid lines mark stable, while dotted lines represent unstable states.

should be chosen accordingly. The regimes of bistability and oscillations were found to be connected producing phase diagrams known in the literature as cross-shaped phase diagrams. Left of the oscillatory domain (OSC) in Fig. 3a the system has only one stable steady state determined by parameters $[S]_0$, pH_0 and E. Moving right, a second stable steady states emerges and the resulting low and high pH steady states are connected via unstable ones (BS). The size of the oscillatory and bistable regimes strongly depended on the amount enzyme encapsulated (Fig. 3b) and P_i whose influence is not discussed herein. Figure 3c presents the evolution of a singe steady state into a bistable structure with increasing enzyme content (from III → I). It should be pointed out that the observation of bistability is largely hindered in experiment by changing $[S]_0$ as high pH states are inaccessible due to the wide span of the bistable regime. A bifurcation diagram complementing Fig. 3a is shown in Fig. 3d.

2.2   Experiments

The successful encapsulation of the enzyme and a pH sensitive probe into the liposomes, is fundamental to match the requisites of differential transport for the acid and the substrate, as devised in numerical simulations. However, the first step in order to obtain the proper experimental conditions, is to test the stability of the enzymatic reaction in the presence of lipids, pyranine and the other ingredients necessary for the encapsulation process.

Thus, the first task was to obtain a pH switch from acid to base in the bulk urea–urease reaction in the presence of lipids and to ensure that no significant side reactions took place. Figure 4 A and B shows the beginning and the end of the reaction in a stirred batch reactor, respectively. Panels C and D show the reaction in the presence of $[POPC] = 0.5 \times 10^{-3}$ M, the colour change indicates that the reaction took place and enough NH_3 was produced to raise the pH.

The reaction dynamics was also monitored in time by following the average red channel intensity of the reaction media in snapshots taken during reactions.

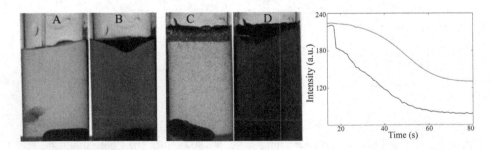

Fig. 4. urea–urease reaction in the absence (panels A and B) and in the presence of $[POPC] = 0.5 \times 10^{-3}$ M (panels C and D). $[urea] = 7.5 \times 10^{-3}$ M, $[H^+] = 5.5 \times 10^{-4}$ M and $[urease] = 16$ units/mL, pH indicator Cresol Red (yellow below pH 7.7, purple above pH 8.8). The right panel shows the kinetics of the urea–urease reaction in the absence (red curve) and in the presence (blue curve) of lipids (Color figure online).

The right panel of Fig. 4 shows the time evolution of the system both in the presence and in the absence of the lipid. Apart for the obvious differences in the intensity values, due to a larger turbidity of the solution in the presence of the lipids, the overall reaction dynamics was quantitatively preserved.

In order to check the response of the fluorescent probe in the same conditions, spectrofluorimetric measurements were conducted in the presence and in the absence of lipids. Even in this case a switch between low and high pH states could be obtained, as showed in Fig. 5, though with a longer clock time compared to the Cresol Red experiments, due to the lower concentration of the enzyme.

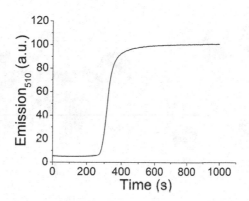

Fig. 5. Spectrofluorimetric trace of the kinetics of the urea–urease reaction. Experimental conditions are the same as in Fig. 4 with [urease] = 10.7 units/mL and [pyranine] = 1 × 10^{-5} M. Excitation λ = 450 nm, Emission λ = 510 nm

Once established the chemical stability of the urea–urease reaction in the presence of lipids and of the fluorescent probe, the next step towards a successful oscillating system was the encapsulation of the enzyme into POPC liposomes through the *droplet transfer method*. This innovative method first takes advantage of the facile compartmentalization of water-soluble solutes (enzyme in this case) in water-in-oil (w/o) droplets, and then convert the solute-filled w/o droplets into vesicles that can be dispersed in an acidic solution of urea.

Figure 6 shows the detailed mechanism: as the first step we prepared the inner solution (I-solution) containing the enzyme ([urease] = 10.3 U/mL) and the fluorescent probe ([pyranine] = 650 μM). Next, we dispersed the I-solution in mineral oil containing [POPC] = 0.5 mM, in order to form the w/o macroemulsion. The lipid formed a monolayer around w/o droplets, stabilising the water/oil interface and preventing phase separation or coalescence (Fig. 6a). A layer of mineral oil containing [POPC] = 0.5 mM was also stratified over the aqueous outer solution (O-solution), in order to prepare an interfacial phase. In this way, a continuous POPC monolayer self-assembles at the oil/water interface (Fig. 6b). Note that I- and O-solutions are isotonic, but their densities (ρ) have been adjusted by adding [sucrose] = 150 mM ($\rho \sim 1.24$ g cm^{-3}) to the I-solution and [glucose] = 150 mM

($\rho \sim$ 1.12 g cm^{-3}) to the O-solution, such as that $\rho_I > \rho_O$. As final step of the preparation, the freshly prepared w/o macroemulsion is poured above the interfacial phase. The emulsion droplets, being denser than oil and denser than O-solution, spontaneously moves across the interface, reaching the O-solution. While crossing the interface, a second POPC layer surrounds the droplets, thus forming a bilayer and consequently the vesicles (Fig. 6c). This step can also be facilitated by centrifuging the sample.

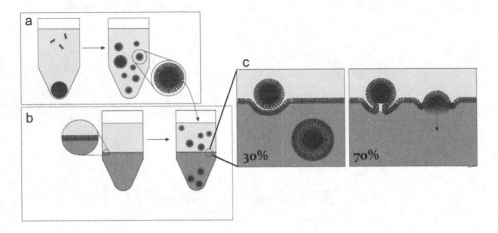

Fig. 6. Sketch of the *droplet transfer method* for vesicles preparation.

Figure 7 (a) and (b) show the result of successful encapsulation of the enzyme and the pH dependent fluorescent probe: panel (a) is the bright field view of panel (b), the latter taken with illumination at $\lambda = 488$ nm, in the proximity of the pyranine absorption maximum. The green colour of the vesicles (probe emits at $\lambda = 511$ nm) indicates that the encapsulation process was successful since pyranine, and most probably all the water-soluble species in the emulsion, were confined inside the double layer of POPC. In this case, pyranine concentration was set ten times larger (650 μM) to enhance the fluorescence intensity of the vesicles. The size of the vesicles was found in the range 5–50 μm of diameter and they were stable over an interval of 2 h, a time that should be sufficient to observe sustained oscillations, as suggested by numerical simulations in Sect. 2.1.

In order to check the presence of the enzyme inside of the vesicles, the liposomes were dispersed in a O-solution containing [urea]$_0$ = 3 × 10^{-3} M, so that a successful permeation of the substrate across the membrane would result in an increasing intensity of the fluorescence. Figure 7 (c) clearly shows the raising intensity over time as a consequence of the increasing pH due to the production of ammonia in the enzymatic reaction inside the vesicles.

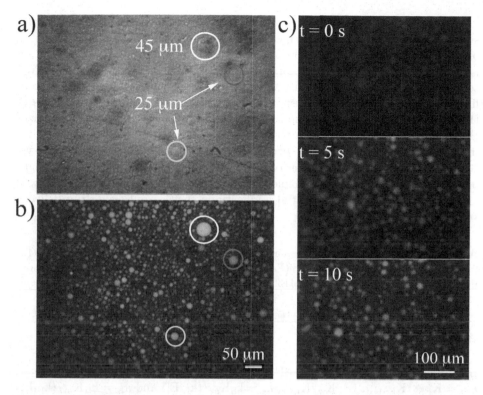

Fig. 7. (a) Bright field view of a sample of vesicles containing [urease] = 10.3 U/mL and [pyranine] = 650 μM. (b) Fluorescence imaging of the sample of vesicles depicted in panel (a) illumination at λ = 488 nm. (c) Fluorescence intensity of the vesicles when immersed in a solution containing $[urea]_0 = 3 \times 10^{-3}$ M, [pyranine] = 650 μM (Color figure online).

3   Conclusions and Perspectives

In this paper, we introduced a new strategy to obtain pH oscillations in a batch reactor by employing the enzymatic urea–urease reaction. In particular, we theoretically demonstrated that, when the reaction is encapsulated in liposomes, the differential transport provided by the different permeability of the reactants through the lipid membrane, can couple with the nonlinear reaction kinetics to generate sustained pH-oscillations. In experiment, we used the *droplet transfer method* to obtain POPC liposomes containing the enzyme and a fluorescent probe sensitive to pH. Preliminary results showed substantial chemical compatibility between the enzymatic reaction and the lipid environment. Switching between low and high pH states was also found in the presence of POPC membranes and a fluorescence probe. Our results are encouraging and the next step will be adding a strong acid to the outer solution, in order to create the negative feedback necessary to lower the pH inside the vesicles, following the autocatalytic production of base in the hydrolysis of urea by urease.

There is increasing interest in feedback in reactions involving biomolecules such as DNA, peptides and enzymes and the spatial and temporal programming of such processes. Positive feedback can result in bistability when the reaction is maintained far from equilibrium, typically negative feedback is then required to generate chemical oscillations. We have shown in simulations how the confinement of the enzyme in a vesicle allows for the separation of timescales of the transport of the feedback species and substrate which in turn destabilizes the steady state providing an alternative route to oscillations. Vesicles play an important role in the storage and transport of biological molecules in living systems. The compartmentalization of enzymes in liposomes may provide insight into the role of confinement of biological catalysts on reaction dynamics as well as creating systems with potential applications such as pulsatile drug delivery *in vivo*.

Acknowledgments. F.R. was supported by the grants ORSA133584 and ORSA149477 funded by the University of Salerno (FARB ex 60 %). The authors acknowledge the support through the COST Action CM1304 (Emergence and Evolution of Complex Chemical Systems).

Appendix

Introducing $s = [S]/K_m$, $h = [H^+]/K_{ES1}$, $\tau = tk_E[E]/K_m$, $\kappa = K_m/K_{ES1}$, $\kappa_{es} = K_{ES2}/K_{ES1}$, $\kappa_w = K_w/K_{ES1}^2$, $\kappa_s = k_S K_m/(k_E[E])$ and $\kappa_h = k_H K_m/(k_E[E])$ Eq. (2) become

$$\frac{\partial s}{\partial \tau} = \kappa_s(s_0 - s) - r$$
$$\frac{\partial h}{\partial \tau} = \left(\kappa_h \left(h_0 - \frac{\kappa_w}{h_0} - h + \frac{\kappa_w}{h} \right) - 2\kappa r \right) \left(1 + \frac{\kappa_w}{h^2} \right)^{-1} \tag{A.1}$$

where

$$r = \frac{s}{(1+s)(1 + \kappa_{es}/h + h)}.$$

References

1. Taylor, A.F.: Mechanism and phenomenology of an oscillating chemical reaction. Prog. React. Kinet. Mech. **27**(4), 247–325 (2002)
2. Orbán, M., Kurin-Csörgei, K., Epstein, I.R.: pH-regulated chemical oscillators. Acc. Chem. Res. **48**(3), 593–601 (2015)
3. Vanag, V.K., Miguez, D.G., Epstein, I.R.: Designing an enzymatic oscillator: bistability and feedback controlled oscillations with glucose oxidase in a continuous flow stirred tank reactor. J. Chem. Phys. **125**, 194515 (2006)
4. Hu, G., Pojman, J.A., Scott, S.K., Wrobel, M.M., Taylor, A.F.: Base-catalyzed feedback in the urea-urease reaction. J. Phys. Chem. B **114**(44), 14059–14063 (2010)

5. Wrobel, M.M., Bánsági, T.Jr, Scott, S.K., Taylor, A.F., Bounds, C.O., Carranza, A., Pojman, J.A.: pH wave-front propagation in the urea-urease reaction. Biophys. J. **103**(3), 610–615 (2012)

6. Stingl, K., Altendorf, K., Bakker, E.P.: Acid survival of Helicobacter pylori: how does urease activity trigger cytoplasmic pH homeostasis? Trends Microbiol. **10**(2), 70–74 (2002)

7. Caplan, S.R., Naparstek, A., Zabusky, N.J.: Chemical oscillations in a membrane. Nature **245**(5425), 364–366 (1973)

8. Turing, A.M.: The chemical basis of morphogenesis. Philos. Trans. R. Soc. Lond. B Biol. Sci. **237**(641), 37–72 (1952)

9. Walde, P., Umakoshi, H., Stano, P., Mavelli, F.: Emergent properties arising from the assembly of amphiphiles. Artificial vesicle membranes as reaction promoters and regulators. Chem. Commun. **50**(71), 10177–10197 (2014)

10. Tomasi, R., Noel, J.M., Zenati, A., Ristori, S., Rossi, F., Cabuil, V., Kanoufi, F., Abou-Hassan, A.: Chemical communication between liposomes encapsulating a chemical oscillatory reaction. Chem. Sci. **5**(5), 1854–1859 (2014)

11. Rossi, F., Zenati, A., Ristori, S., Noel, J.M., Cabuil, V., Kanoufi, F., Abou-Hassan, A.: Activatory coupling among oscillating droplets produced in microfluidic based devices. Int. J. Unconv. Comput. **11**(1), 23–36 (2015)

12. Torbensen, K., Rossi, F., Pantani, O.L., Ristori, S., Abou-Hassan, A.: Interaction of the Belousov-Zhabotinsky reaction with phospholipid engineered membranes. J. Phys. Chem. B **119**(32), 10224–10230 (2015)

13. Stockmann, T.J., Nol, J.M., Ristori, S., Combellas, C., Abou-Hassan, A., Rossi, F., Kanoufi, F.: Scanning electrochemical microscopy of Belousov-Zhabotinsky reaction: how confined oscillations reveal short lived radicals and auto-catalytic species. Anal. Chem. **87**(19), 9621–9630 (2015)

14. Vanag, V.K., Epstein, I.R.: Pattern formation in a tunable medium: the Belousov-Zhabotinsky reaction in an aerosol OT microemulsion. Phys. Rev. Lett. **87**(22), 228301–228304 (2001)

15. Toiya, M., Vanag, V.K., Epstein, I.R.: Diffusively coupled chemical oscillators in a microfluidic assembly. Angew. Chem. Int. Ed. **47**(40), 7753–7755 (2008)

16. Rossi, F., Vanag, V.K., Epstein, I.R.: Pentanary cross-diffusion in water-in-oil microemulsions loaded with two components of the Belousov-Zhabotinsky reaction. Chem. Eur. J. **17**(7), 2138–2145 (2011)

17. Tompkins, N., Li, N., Girabawe, C., Heymann, M., Ermentrout, G.B., Epstein, I.R., Fraden, S.: Testing turings theory of morphogenesis in chemical cells. Proc. Nat. Acad. Sci. **111**(12), 4397–4402 (2014)

18. Paul, A.: Observations of the effect of anionic, cationic, neutral, and zwitterionic surfactants on the Belousov-Zhabotinsky reaction. J. Phys. Chem. B **109**(19), 9639–9644 (2005)

19. Rossi, F., Varsalona, R., Liveri, M.L.T.: New features in the dynamics of a ferroin-catalyzed Belousov-Zhabotinsky reaction induced by a zwitterionic surfactant. Chem. Phys. Lett. **463**(4–6), 378–382 (2008)

20. Jahan, R.A., Suzuki, K., Mahara, H., Nishimura, S., Iwatsubo, T., Kaminaga, A., Yamamoto, Y., Yamaguchi, T.: Perturbation mechanism and phase transition of AOT aggregates in the Fe(II)[batho(SO3)2]3 - catalyzed aqueous Belousov-Zhabotinsky reaction. Chem. Phys. Lett. **485**(4–6), 304–308 (2010)

21. Rossi, F., Liveri, M.L.T.: Chemical self-organization in self-assembling biomimetic systems. Ecol. Modell. **220**(16), 1857–1864 (2009)

22. Sciascia, L., Rossi, F., Sbriziolo, C., Liveri, M.L.T., Varsalona, R.: Oscillatory dynamics of the Belousov-Zhabotinsky system in the presence of a self-assembling nonionic polymer. Role of the reactants concentration. Phys. Chem. Chem. Phys. **12**(37), 11674–11682 (2010)

23. Rossi, F., Varsalona, R., Marchettini, N., Turco Liveri, M.L.: Control of spontaneous spiral formation in a zwitterionic micellar medium. Soft Matter **7**, 9498 (2011)

24. Rossi, F., Budroni, M.A., Marchettini, N., Carballido-Landeira, J.: Segmented waves in a reaction-diffusion-convection system. Chaos Interdisc. J. Nonlinear Sci. **22**(3), 037109-1–037109-11 (2012)

25. Budroni, M.A., Rossi, F.: A novel mechanism for in situ nucleation of spirals controlled by the interplay between phase fronts and reaction diffusion waves in an oscillatory medium. J. Phys. Chem. C **119**(17), 9411–9417 (2015)

26. Taylor, A.F., Tinsley, M.R., Wang, F., Huang, Z., Showalter, K.: Dynamical quorum sensing and synchronization in large populations of chemical oscillators. Science **323**(5914), 614–617 (2009)

27. Rossi, F., Ristori, S., Marchettini, N., Pantani, O.L.: Functionalized clay microparticles as catalysts for chemical oscillators. J. Phys. Chem. C **118**(42), 24389–24396 (2014)

28. Bánsági, T., Taylor, A.F.: The role of differential transport in an oscillatory enzyme reaction. J. Phys. Chem. B **118**(23), 6092–6097 (2014)

29. Muzika, F., Bansagi, T., Schreiber, I., Schreiberovi, L., Taylor, A.F.: A bistable switch in pH in urease-loaded alginate beads. Chem. Commun. (Camb.) **50**(76), 11107–11109 (2014)

30. Lasic, D.D., Barenholz, Y.: Handbook of Nonmedical Applications of Liposomes: Theory and Basic Sciences, vol. 1. CRC Press, Boca Raton (1996)

31. Paula, S., Volkov, A., Van Hoek, A., Haines, T., Deamer, D.W.: Permeation of protons, potassium ions, and small polar molecules through phospholipid bilayers as a function of membrane thickness. Biophys. J. **70**(1), 339 (1996)

32. Pautot, S., Frisken, B.J., Weitz, D.A.: Production of unilamellar vesicles using an inverted emulsion. Langmuir **19**(7), 2870–2879 (2003)

33. Carrara, P., Stano, P., Luisi, P.L.: Giant vesicles colonies: a model for primitive cell communities. ChemBioChem **13**(10), 1497–1502 (2012)

34. Stano, P., Souza, T.P.D., Carrara, P., Altamura, E., DAguanno, E., Caputo, M., Luisi, P.L., Mavelli, F.: Recent biophysical issues about the preparation of solute-filled lipid vesicles. Mech. Adv. Mater. Struct. **22**(9), 748–759 (2015)

35. Stano, P., Wodlei, F., Carrara, P., Ristori, S., Marchettini, N., Rossi, F.: Approaches to molecular communication between synthetic compartments based on encapsulated chemical oscillators. In: Pizzuti, C., Spezzano, G. (eds.) WIVACE 2014. CCIS, vol. 445, pp. 58–74. Springer, Heidelberg (2014)

36. Grotzky, A., Altamura, E., Adamcik, J., Carrara, P., Stano, P., Mavelli, F., Nauser, T., Mezzenga, R., Schlter, A.D., Walde, P.: Structure and enzymatic properties of molecular dendronized polymer-enzyme conjugates and their entrapment inside giant vesicles. Langmuir **29**(34), 10831–10840 (2013)

37. Clement, N.R., Gould, J.M.: Pyranine (8-hydroxy-1,3,6-pyrenetrisulfonate) as a probe of internal aqueous hydrogen ion concentration in phospholipid vesicles. Biochemistry **20**(6), 1534–1538 (1981)

38. Ermentrout, B.: Simulating, Analyzing, and Animating Dynamical Systems: A Guide to XPPAUT for Researchers and Students, vol. 14. Siam, Philadelphia (2002)

Author Index

Printed in the United States
By Bookmasters